AF701497

Blown Flathead

How to Build a Supercharged Flathead Ford for the Street

by
Joseph P. Abbin, P.E.

Roadrunner Engineering
Albuquerque, New Mexico

Copyright ©2000, 1998 by Joseph P. Abbin, P.E.
All Rights Reserved

Printed in the United States of America
No part of this book may be reproduced without written permission from the author, except for brief quotations which are properly credited.

All information, opinions and data in this book have been thoroughly researched by the author. There is no guarantee as to the accuracy or completeness of the material presented. The reader assumes any and all risks associated with the use of the contents of this book.

Layout, design and editing by
Kathleen McCaughey and Celene Morningstar.
Roadrunner logo and Rowdy Roadrunner artwork
by Corey Weintraub.

Published by
Motorhead Mart
PO Box 53296
Albuquerque, New Mexico 87153
(505) 268-6768

Printed by
Aiken Printing Co.
1112 Third Street NW
Albuquerque, New Mexico 87102
(505) 243-5631

Second Edition
ISBN 1-880047-81-0

Dedication

This book is dedicated to flathead Ford fans
who want to run with the big dogs and
move to the head of the pack.

Flatheads Forever!

Acknowledgements

Much of the material presented in this book was gleaned from a variety of written sources and personal communications with many "flathead" people. I have tried to properly credit these sources. I apologize for any omissions and thank all those who contributed to this book.

About the Author

Joe Abbin is the owner of Roadrunner Engineering and the Motorhead Mart in Albuquerque, New Mexico. Roadrunner Engineering specializes in consulting for high performance automotive engines, including the flathead Ford. Roadrunner Engineering also produces flathead Ford supercharger kits. Joe has a Masters degree in Mechanical Engineering and is a registered professional engineer.

Joe has over forty years of practical, hands-on automotive experience as well as analytical credentials. He is the owner and engine builder for the "Motorhead Mart Special" – a well-known blown flathead powered nostalgia racer in the Southwest. The Motorhead Mart is an auto-related bookstore which currently specializes in books on the flathead Ford engine.

Foreword

In 1960, I pulled the faithful flathead out of my 1940 Ford coupe. In it's place, I dropped in a 1956 Mercury Y-block engine. The old flathead had a 3-3/8" bore by 4" stroke, Navarro high compression heads, Isky 400 Jr. cam, Kong ignition, Edelbrock manifold with three 2-barrel Holley 94's, headers and other goodies. I traded the flathead for a stock 1934 Ford Tudor sedan to drive while I worked on the 1940! Thirty plus years later, I got an unexplained case of yearning (i.e. nostalgia) to build another flathead. . . I bought a derelict 1934 Ford Tudor which had been resting after an early 60's career as a drag racer with a 327 cubic inch Chevy engine. I rebuilt it, installed a 1949 Mercury flathead and called it the "Motorhead Mart Special." The "Special" far outperforms my old flathead-powered 1940 coupe. Wherever it goes, it makes friends who often have never seen a flathead engine in an old Ford! It's performance commands respect from it's overhead-valve street rod brothers. The supercharger combination is the key and this book offers guidance on how to put such a combination together and allow your flathead to kick some asphalt.

Joe Abbin

ROADRUNNER

Table of Contents

CHAPTER 5
BUILDING A COMPLETE PACKAGE

APPENDICES

1

INTRODUCTION

This chapter discusses the goals and methods used in producing this book.

PURPOSE OF THIS BOOK

The sole purpose of this book is to provide a "how-to recipe" for building a high performance ***street*** flathead Ford or Mercury engine with readily available parts. High performance is relative, but

Figure 1-1: S.CO.T. Supercharger installed in a 1933 Ford. This is a Roots-type blower marketed in the early 1950's. What a beauty!

the guidelines presented in this book will enable the flathead powered car owners to run with their overhead valve powered neighbors and surprise more than a few without breaking either the engine or the wallet.

WHY BUILD A FLATHEAD FOR A "MODERN" STREETROD? THE PROS AND CONS

Why build a wooden boat? Why fix a vacuum tube radio? Why build a fabric-covered airplane? Indeed. Why not use the latest and best engine technology we can afford? These are easy questions to answer but hard to explain. The simple answer is we want to! Building a 40-something year old flathead is not cheap and despite our efforts and optimism it probably won't blow off a similarly built overhead valve engine.

Nevertheless, there is a certain satisfaction in doubling or tripling the horsepower of one of these old engines and giving the overheads a run for their money. A high performance flathead looks and sounds great. They are the best fit in an old Ford cuz that's the way Henry designed them from 1932 to 1953. And, contrary to what many street rodders think, a high performing flathead can be built to run coolly and reliably. That's what this book is all about - so let's get on with it!

COMPARISON WITH OTHER ENGINES

Flathead Fords and Mercs are comparable in size and weight to small block overhead valve Fords (e.g. 351 cubic inch) and Chevys (e.g. 350 cubic inch). They all weigh approximately 550 pounds. Big block Ford (e.g. 460 cubic inch) and Chevys (e.g. 454 cubic inch) weigh approximately 700 pounds and are significantly bigger. Early Chrysler hemis (e.g. 392 cubic inch) are justifiably nicknamed "elephants" weighing in at about 800 pounds and are dimensionally the biggest commonly swapped engine. The overheads get about 0.5-0.7 hp/cubic inch stock versus about 0.4-0.5 hp/cubic inch for the stock flatheads so we have a "performance deficit" to overcome.

MODERN METHODS FOR ENHANCING FLATHEAD PERFORMANCE

Note that the word "methods" is used above versus "equipment." All of the equipment to provide a high performing flathead has been available in one form or another since 1959 and earlier.

Two major breakthroughs in recent years have contributed to major internal combustion engine performance improvements. These are the now widespread availability and use of the flow bench to study engine airflow, and computer simulations to evaluate the infinite number of engine "combinations." These tools have allowed rapid advances in automotive engine performance. Flow bench results have been very closely correlated to actual engine performance and are much easier, quicker, and cheaper to obtain than dynamometer or racetrack performance. "If it don't flow, it won't go" is a proven statement.

Chapter 3 and Appendix A of this book present flow test results on different flathead combinations from tests performed by the author and Ron Meek of Flow Dynamics in Albuquerque, New Mexico. Much of the data validates the work done by the old time flathead hot rodders, but there are fresh insights into the relative importance of different modifications and the tradeoff between compression ratio versus volumetric efficiency of the flathead.

Computer simulations of internal combustion engine performance were very crude until recent years. Indeed, predicting engine performance was much like predicting the weather or the stock market. Today, there are powerful software tools available for simulating automotive engines including the flathead. My company, Roadrunner Engineering, both uses and is participating in the development of this software. This software allows flow test data to be input for the simulations–thus greatly improving the accuracy for a specific engine. All these simulations should be correlated and "baselined" as necessary with actual dyno and/or track data.

Assuming this is done, other engine combinations can be compared with the baseline quite readily and at very low cost, certainly much cheaper than the "build and test" method. Most of the flathead performance curves presented in this book are from computer simulations utilizing actual data for calibration of the codes. Evaluating a camshaft change, for example, becomes a simple matter of inputting the cam specifications and pushing a button. Ah, technology is wonderful at times.

The author's own car does not utilize any real high tech equipment, however the combination has been carefully simulated with high tech software to get a cost-effective, high-performance flathead for the street. This would not have been possible even ten years ago.

Flathead engine analyses are a specialty of my company, Roadrunner Engineering. A full analysis is included in the purchase price of flathead supercharger kits sold by Roadrunner. For further information on flathead engine analysis and supercharger kits, call, write or e-mail me at:

Roadrunner Engineering

P.O. Box 53296

Albuquerque, NM 87153-3296

phone/fax (505) 268-6768

or e-Mail: RoadrunnerEngr@msn.com

2

FUNDAMENTALS

This chapter discusses the flathead family tree and general tips for building a cool running, reliable flathead.

EVOLUTION OF THE FLATHEAD

Figure 2-1: 1932 Ford Cabriolet and Roadster These cars were a hit from the very beginning. Henry Ford's V-8 beauties were the "Official Cars" for the 1932 Indianapolis 500 Race. Famous flyer Capt. Eddie Rickenbacker is at the wheel of the Cabriolet on the left. Several flathead Fords actually competed in the Indy 500 races over the years.

Henry Ford introduced the first Ford flathead V-8 in the 1932 models to counter Chevrolet's overhead valve six, which was introduced in 1929 and contributed greatly to Chevrolet's inroads on Ford sales. The 1932 Ford V-8 was quite a sensation with a whopping 65 horsepower! There were overheating problems primarily due to the approximately 3 feet of exhaust passages in the block and the placement of the water pumps up high on the head that often cavitated when hot. The water pump problem was finally fixed in 1937 by moving them down low on the block. There were also problems with high oil consumption which was rectified with four ring pistons and better rings in 1933.

Vapor lock was also a problem due to the high (and hot) placement of the fuel pumps on top of the back of the engine. The fuel pump design was improved, but the placement was never fixed and vapor lock is a flathead trademark with the original fuel pumps.

The horsepower was increased to 75 horsepower in 1933 with an increase in compression ratio from 5.5:1 to 6.3:1. The horsepower took another jump to 85 horsepower in 1934 with the addition of a 2-barrel carburetor.

The 1934 Fords are probably the hottest performing stock flathead powered cars due to their combination of relatively high horsepower and light weight. This performance made them a favorite of stock car racers, moonshiners and gangsters of the day. Clyde Barrow of "Bonnie and Clyde" fame even went so far as to write a letter to Henry Ford praising the virtues of the Ford as a great getaway car. All flathead Fords and Mercs were considered snappy performers up through 1941. The increases in horsepower did not keep up with the increases in weight from 1942 onward, although the 1952 and 1953 Mercury flatheads were rated at 125 horsepower. Table 2-1 summarizes the flathead evolution from 1938 to 1953.

IDENTIFYING AND CHOOSING THE BASIC ENGINE

The blocks built from July 1938, to the end of production in 1953, all had 24 studs or cap screws retaining the heads. These are the only blocks which I would consider for performance applications. The 1937 and earlier blocks should be reserved for collector cars and speed equipment availability is limited anyway. The 1938-1942 Ford passenger car engines displace 221 cubic inches and have a 3-1/16 inch bore and a 3-3/4 inch stroke (3-1/16 x 3-3/4). The 1939-1948 Mercury and 1946-1953 Ford passenger car and light truck engines displace 239 cubic inches (3-3/16 x 3-3/4). The 1949-1953 Mercury engines displace 255 cubic inches (3-3/16 x 4). The listing above is in order of increasing desirability for performance application.

The later engines (1949-1953), both Ford and Mercury, have the largest displacements, modern bearing and distributor designs, as well as improved oiling and cooling systems.

The major specifications for the 1938-1953 flatheads are given in Table 2-1. Identifying blocks can often be done most reliably by actually measuring key dimensions. Flathead Readings 2 and 3 in Appendix F include information on block identification.

CLEANING AND CHECKING THE BLOCK

Once you have located and identified a candidate block, it must be thoroughly cleaned and inspected for useability. A conventional hot tanking in a caustic solution is a minimum requirement. Acid dipping is desirable after an initial caustic clean when significant rust is present in the cooling passages. A professional acid cleaning of the block will effectively remove the rust without attacking the parent metal. There are commercial firms which can do the acid cleaning. If there is not such a facility readily available, it can be done by the builder. (See Appendix G)

After cleaning, the block can be inspected for cracks and other flaws by Magnafluxing or dye penetrant. Cracks in the valve pocket area, in the cylinders or in the center main bearing crankshaft support web, are cause for rejection. Cracks between the head stud holes in the block and the adjacent cooling holes are common, but are not a problem and no repair is required. Cracks in other areas, such as along the oil pan rail may be successfully repaired and should be evaluated by a competent engine builder.

Year	Ford (F) Merc (M)	Model	Bore	Stroke	Disp	Comp	Rated HP	Torque	Comments
1938	F	81A	3-1/16	3-3/4	221	6.12	85@3800	146@2000	First 24 stud heads
1939	F	91A	3-1/16	3-3/4	221	6.15	90@3800	155@2200	
1939	M	99A	3-3/16	3-3/4	239	6.30	95@3800	170@2100	
1940	F	01A	3-1/16	3-3/4	221	6.15	90@3800		
1940	M	09A	3-3/16	3-3/4	239	6.30	95@3800		Same as 1939
1941	F	11A	3-1/16	3-3/4	221	6.15	90@3800		
1941	M	19A	3-3/16	3-3/4	239	6.30	95@3800		Same as 1939
1942	F	21A	3-1/16	3-3/4	221	6.20	96@3800	155@2200	
1942	M	19A	3-3/16	3-3/4	239	6.40	100@3800	180@2000	
1946	F	69A	3-3/16	3-3/4	239	6.75	100@3800	180@2000	
1946	M	69M	3-3/16	3-3/4	239	6.75	100@3800	180@2000	
1947	F	79A	3-3/16	3-3/4	239	6.75	100@3800		Same as 1946
1947	M	79M	3-3/16	3-3/4	239	6.75	100@3800		
1948	F	89A	3-3/16	3-3/4	239	6.75	100@3800		Same as 1946
1948	M	89M	3-3/16	3-3/4	239	6.75	100@3800		
1949	F	8BA	3-3/16	3-3/4	239	6.80	100@3600	180@2000	
1949	M	9CM	3-3/16	4	255	6.80	110@3600	200@2000	
1950	F	0BA	3-3/16	3-3/4	239	6.80	100@3600	181@2000	
1950	M	0M	3-3/16	4	255	6.80	110@3600	200@2000	
1951	F	1BA	3-3/16	3-3/4	239	6.80	100@3600	187@1800	
1951	M	1M	3-3/16	4	255	6.81	112@3600	206@2000	
1952	F	B2	3-3/16	3-3/4	239	7.20	110@3800	194@1900	
1952	M	BC	3-3/16	4	255	7.20	125@3700	218@1700	
1953	F	B3	3-3/16	3-3/4	239	7.20	110@3800	196@2000	
1953	M	BG	3-3/16	4	255	7.20	125@3800	218@1700	

Table 2-1: Partial Overview of Ford and Mercury Flathead V-8 Specifications

REBUILDING THE FLATHEAD

A flathead can be rebuilt by any competent auto machine shop, although it is desirable to find a shop which has prior experience and hopefully an interest. No out-of-the-ordinary equipment is required, although special tools, such as a bar-type valve spring compressor is handy. Parts are readily available, although maybe not locally.

Prices on parts and labor are comparable to equivalent quality items for a modern overhead valve V-8 engine. The home builder will not be over challenged by the task of rebuilding a flathead. Any of the service manuals and several of the references listed in Appendix F will be excellent reading for the rebuilder.

A list of potential suppliers for flathead services and parts are listed in Appendix G.

This book will concentrate on performance modifications for the flathead. Standard rebuilding techniques and common modifications are covered very well in the references. The only item which I feel needs to be stressed for any flathead rebuild is the necessity to thoroughly clean the block. As pointed out by other authors and reinforced by my own experience, time spent getting all the cooling passages clean is time well spent. Done properly, you will have a dependable, cool running flathead. As noted previously, there are almost three feet of exhaust passages running through the block, which certainly stresses the cooling system, requiring the entire system to be top notch for proper cooling. In cleaning my own blocks, I spend considerable time manually probing and prying rust, core sand (yes, left over from the factory), other "dirt", etc. out of the cooling passages with a coat hanger wire, screwdriver, etc. and rolling the block on soft ground and flushing with a hose prior to hot tanking the block. You will be amazed at what comes out!

FLATHEAD COOLING TIPS

Assuming the block is clean and the rest of the cooling system is in good shape, there are several other measures we can take to ensure a cool running flathead when we double or even triple the horsepower output. These are listed below.

1. Use a pressurized cooling system. Four to seven lbs/in^2 is generally safe for even the early systems if the radiator is in good shape. The big radiator tanks on the pre-1938 cars can not be safely used with higher pressures without risking the tanks separating from the core. The pressure on the system helps the water pumps work more effectively by minimizing cavitation (vapor formation) at the inlets when hot. The pressurized system also helps prevent vapor formation or boiling in the heads or block and thus also prevents hot spots where the vapor bubbles form. Obviously, pressurization also helps prevent coolant loss through the overflow system due to boiling or even water pump pressure.

Figure 2-2: Flathead Water Pump Impellers
Left - Pre 1938; Middle - 1938-1948 (1938-1947 Truck)
Right - 1949-1953 (1948-1953 Truck)

Figure 2-3: 1938-1948 Water Pump (left) and
1949-1953 (1948-1953 Truck) Water Pump (right)

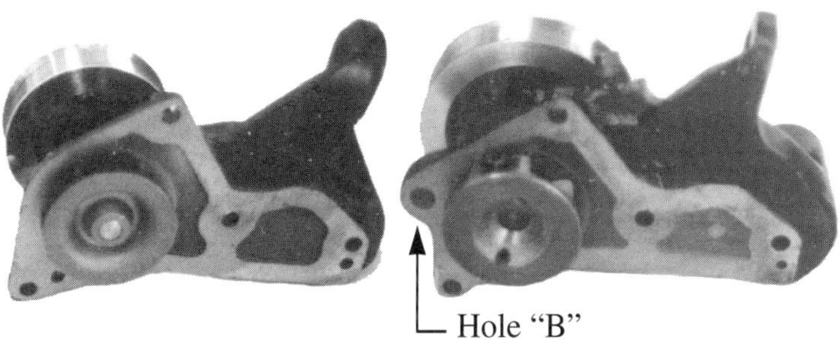

Figure 2-4: 1938-1948 Water Pump (left) and
1949-1953 (1948-1953 Truck) Water Pump (right)

2. Use later style water pumps. The first flathead V-8 water pumps had only 3 or 4 blades or vanes. In the midyears they had 6 vanes, and finally 8 vanes on the later passenger cars and trucks.

   Besides the additional number of vanes, the vanes on the later impellers are taller, larger in diameter and curved to enhance the flow. The later pumps are currently being manufactured new and can be used in the older blocks in many applications by trimming the impellers (reducing the height 0.1" and the diameter 0.070") and blocking the bypass port (Hole "B" in Figure 2-4) on the top of the pump.

   These new pumps also have modern seals and permanently lubricated bearings. The increased pump pressure and flow enhances cooling substantially. Putting restrictors or thermostats in the water outlets in the heads can also enhance cooling by increasing the water pressure in the heads and block, thereby further discouraging boiling and hot spots. Removing vanes does not make sense except for sustained high speed engines with excess pump capacity (no street flathead engine falls in this category).

3. Tune the engine. An ignition which is retarded or over-advanced produces less horsepower and more heat. Likewise, a lean fuel/air mixture (jets too small, vacuum leak, etc.) causes the engine to run hot. See Chapter 5 for tuning tips.

4. Use a fan. Either an electric or a mechanical (if space allows) fan is a necessity for urban street driving.
5. Make sure air can flow. Ensure a suitable grill opening, louvers in the hood and side panels, shrouds and air ducts as appropriate. Some after market radiators have too many fins per inch and do not allow sufficient airflow, i.e. the air would rather flow around rather than thru the radiator. Seven or eight fins per inch like the original Ford seems to work well even with water tubes four rows deep.
6. Make sure water can flow. Ensure that the radiator is clean and there are no kinks or obstructions in the hoses. Lower hoses should have internal springs or be sufficiently stiff to prevent collapse when hot. Verify function of the thermostats if used. A 0.125" hole drilled in the non-movable part of the thermostat assists proper filling and helps ensure water movement and proper operation of the thermostat as the water heats up.

3

HOW TO MAKE IT GO

This chapter discusses traditional hop up techniques and trade-offs for the flathead.

Figure 3-1: Normally Aspirated "Full Dress" Street Flathead

THE TRIED AND TRUE METHODS

Engine horsepower is proportional to engine torque multiplied by rotational speed. Thus, we can increase horsepower by increasing torque and/or speed. There are many ways we can do this, however there are limits to what we can achieve in each area. Increasing the engine speed of a street flathead with mostly stock parts to much above 5000 rpm is not practical because of structural considerations as well as breathing limitations. Therefore, we will concentrate on torque.

For increasing torque and horsepower, you've probably heard the old saying, "there's no substitute for cubic inches." For now, we'll accept this since all else being equal, the torque and horsepower go up directly with cubic inches or displacement. In the racing world, this advice has been taken to heart with "mountain" motors of over 800 cubic inches running in unlimited classes. In the street rod world, there are big blocks of all brands running 440 cubic inches or more. Even the common 350 cubic inch Chevys and 351 cubic inch Fords are bigger than any "mountain motor" flatheads. In the old days, and even now, 296 cubic inches was/is a "big" flathead. Due to the scarcity of good blocks and crankshafts, most current flathead builders limit the displacement to 276 cubic inches or less to allow room for future rebuilds. This is hardly a mountain motor, but still a respectable base for good performance.

Thus if we can take a 239 cubic inch flathead Ford and bore it 1/8 inch to 3-5/16 inch and install a '49 - '53 Mercury crank with a 4 inch stroke, we get about 276 cubic inches. This 15% increase in displacement results in an equivalent horsepower gain.

Other popular hopups for the flathead are headers and dual exhausts, dual carburetors, a hotter cam, porting and relieving, and higher compression heads.

#	Combinations	Torque (Ft-lbs)	Peak HP	Approx. Cost	$/HP
1	Stock '48 Merc - factory dyno with fan & muffler*	180 @ 2000	90 @ 3600		
2	#1 + Dual Exhausts & Headers	190 @ 2000	100 @ 3500	$ 250	$25
3	#2 + Dual Carbs or 390 cfm 4-bbl	195 @ 2000	110 @ 3500	$ 500	$50
4	#3 + "3/4 Race" Street Cam & Kit	190 @ 2500	130 @ 4000	$ 400	$20
5	#4 + Overbore 0.125" with 4" Crank	235 @ 2500	150 @ 4000	$ 1100	$55
6	#5 + Port & Relieve with 1.6" Valves	235 @ 2500	175 @ 4500	$ 900	$35
7	#6 + Aluminum Hi-Comp Heads (63cc)	255 @ 2500	180 @ 4500	$ 400	
8	#5 + 6-7 psi Supercharger with 600 cfm 4-bbl	280 @ 2500	195 @ 4500	$ 2400	$55
9	#6 + 6-7 psi Supercharger with 600 cfm 4-bbl	305 @ 2500	260 @ 5000	$ 2400	$30
10	1981 Chevrolet 350 Z28, 4-bbl (reference)	275 @ 2400	175 @ 4000		
11	1981 Ford 351 High Output (reference)	285 @ 2200	165 @ 3600		

** Factory rated HP in Table 2-1 is higher but does not include the fan or muffler.*

Table 3-1: Flathead Ford/Mercury Buildup Combinations Versus Torque (Ft-lbs) and Peak Horsepower (both at Sea Level, 60 deg F, dry air)

Table 3-1 shows horsepower versus rpm for a stock flathead with a typical progression of common hopups as listed above. Figure 3-2 illustrates the horsepower versus rpm curves for the stock '48 Mercury flathead, a normally aspirated "high performance street" flathead (combination #7 in Table 3-1) and a supercharged "high performance street" flathead (combination #9). The differences are dramatic! There's life in that old flathead to be sure! It should be noted that the relative power increases for each individual hopup or modification would be different if the modifications are done in a different order. Also, there can be a significant difference in the degree of each modification, such as the amount of porting and relieving, the blower boost, etc., resulting in a significant difference in the horsepower increase at each step.

Also shown in Table 3-1 are the approximate costs of the modifications for comparison. Dual exhausts, a hotter cam, and porting and relieving are quite cost effective. The modification costs can be more or less depending on if the mods are done separately or in combination.

As you can see from Table 3-1 and Figure 3-2, a strong naturally aspirated street flathead will put out about 150-180 honest horsepower in the vehicle. For many folks, a 50-100% increase in torque and horsepower, an engine that looks and sounds good and yet is still reliable and cool running is enough. In a light car, such as a 1932 roadster, the driver can expect 0-60 mph times of 6 seconds and quarter mile times of 15 seconds, with quarter mile speeds of 90 mph at sea level. One can also expect to run competitively with a similar car powered by a stock overhead valve 350 cubic inch Chevy or 351 cubic inch Ford. For me, that's not good enough, so we will keep going!

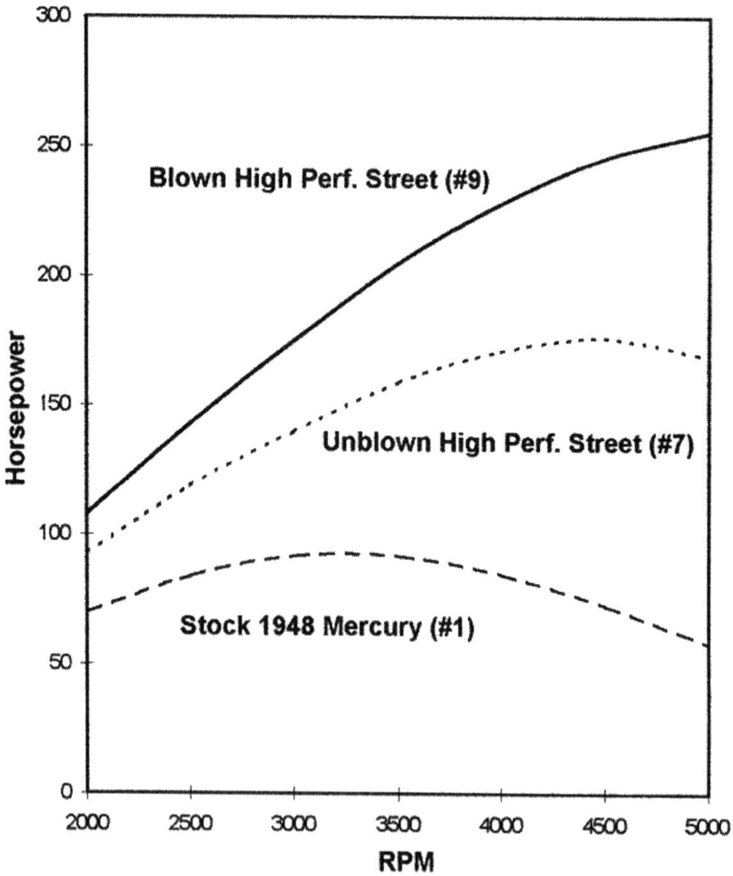

Figure 3-2: Horsepower Versus RPM for Three Combinations of Flathead Modifications #1, #7, and #9 from Table 3-1

THE TRADE OFF BETWEEN COMPRESSION RATIO AND BREATHING

The flathead has a much more severe problem in the breathing area than an overhead valve engine, and unfortunately, almost everything we do in a flathead combustion chamber to improve breathing hurts compression ratio. Figure 3-3 shows a cross-section of the flathead combustion chamber and intake system in the block.

Referring to Figure 3-3, we note that the flathead intake and exhaust flow must travel laterally from the valves to the cylinder and vice versa. This causes a major inefficiency in the flathead breathing capability although it does promote good "swirl" and "tumble" of the flow and thus enhances fuel mixing and combustion.

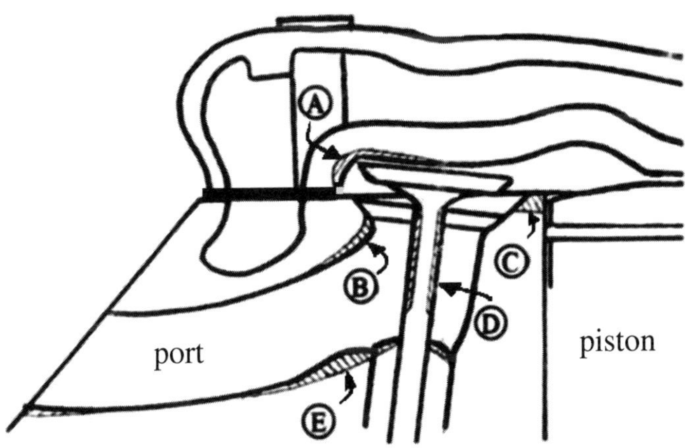

Figure 3-3: Cross-Section of the Flathead Combustion Chamber and Intake System

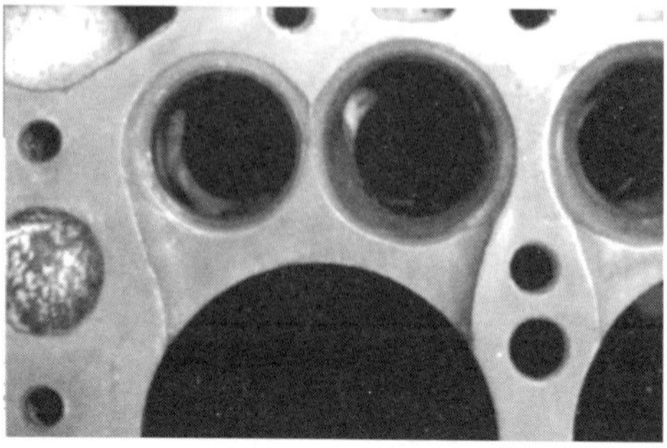

Figure 3-4: Block Relief

Relieving the block, i.e. removing metal from the block at "C" in Figure 3-3, and as shown in Figure 3-4, enhances flow. Opening up the head by removing metal as shown at "A" in Figure 3-3, also enhances flow particularly at low to mid valve lifts. However, both these modifications increase the combustion chamber volume and lower the compression ratio, which in turn lowers horsepower and fuel economy. Most traditional high compression heads for the flathead raise the compression ratio by minimizing the combustion chamber volume including the flow passage area between the valves and the piston which restricts flow. In general, the higher the compression ratio of the heads, the more restricted is the breathing, no matter what brand. Thus we have a trade-off between performance enhancers, and we must be able to measure or estimate the gains from each option to arrive at an optimum. To do this, we must measure or estimate the compression ratio and the flow (breathing) characteristics of the engine.

To evaluate the effects of different modifications on flow, Appendix A contains a small part of the flow test data generated in the author's study of the flathead. An entire book could easily be written on the subject of porting or otherwise enhancing the intake and exhaust flow of the flathead and the effects on performance.

While it is difficult to generalize the results of all this modifying and flow testing, Tables 3-2 and 3-3 give estimates of the flow changes to be expected with different modifications to the intake and exhaust systems. The results presented are representative of those which can be obtained on a street engine. Competition engines with filled water passages can have greatly different results since breakout to the cooling passages is not a concern.

Modification See Figure 3-3	**Effects on Intake Flow**	**Comments**
Undercutting head to gasket boundary, "A"	+ 2-7%	Most benefit low to mid valve lifts
Rolling short side intake port radius, "B"	+ 4-7%	
Relieving block, "C"	+ 5-15%	High compression head with 63cc combustion chamber benefited 15%
Porting block, "E"	+ 10-20%	Opening port, minimizing valve guide obstruction and widening port in turn area
High flow valves, "D"	+ 1-3%	Underside of valve and upper part of stem undercut
1.6" diameter valve	+ 5-10%	Most benefit @ low to mid lifts
1.72" diameter valve	—	No benefit without excessive head modifications
High compression head 63cc combustion chamber	- (4-12)%	versus undercut stock

Table 3-2: Summary of Flow Changes Versus Intake System Modifications

Modification See Figure 3-3	Effects on Flow		Comments
	Center Exhaust	End Exhaust	
Undercut head, "A"	+ 5-10%	+ 3-8%	
Relieving block, "C"	+ 6-13%	+ 1-2%	Most benefit at high lifts
Smooth outlet port	—	+ 10-15%	End exhaust benefits greatly
Porting block, "E"	+ 4-5%	+ 10-25%	Opening bowl port very effective on end exhausts
High flow valves, "D"	+ 2-3%	+ 2-5%	
1.6" diameter valve	no test	+ 5-10%	Most benefit @ low to mid lifts
High compression head (63cc)	<1%	<1%	versus undercut stock

Table 3-3: Summary of Flow Changes Versus Exhaust System Modifications

Referring to Table 3-2 or Appendix A, we note that indeed the 63cc high compression aluminum head does not breathe (flow) as well as a stock head. To determine if such heads increase horsepower or not, we will estimate both flow and compression ratio effects on power for this specific head.

Horsepower is proportional to flow, all else being equal. Thus we would expect a corresponding **decrease** in peak horsepower of 4-12% due to flow considerations alone using the 63cc head. If the high compression heads replace stock 8BA heads and increase compression ratio from 7.5:1 (relieved) to about 8.6:1, as described in two examples in Appendix B, we could expect a horsepower **increase** of about 10-15% due to compression ratio alone.

Combination #7 in Table 3-1 shows the results of analyzing both of the above effects together. We see that torque and horsepower increase about 9% at 2500 rpm, but the benefit dwindles to about 3% over stock heads at 4500 rpm. Therefore, the use of most high compression heads on a flathead results in very modest gains in **peak** horsepower but can significantly improve low end torque. For an unblown street flathead, I recommend a compression ratio of 8.5:1 as a reasonable compromise between low and high rpm power and fuel economy. Fuel octane and structural limitations suggest compression ratios about one point lower (ie. 7.5:1) for blown flatheads as discussed in the following chapter.

Referring to another example in Appendix B, where we look at the effects of relieving a 3-5/16 x 4 engine with stock heads, we find that the compression ratio is reduced from 7.7:1 to 7.5:1. If all other things were equal, lowering the compression ratio 0.2 would hurt horsepower about 1%. However, we know that relieving increases flow about 5-15% (from Table 3-2). Thus, relieving does represent an overall performance gain and is almost mandatory for traditional high compression heads. Partial relieving, as recommended by some builders, where the relief does not continue all the way to the cylinder bore, was found to flow less than no relief.

4

SUPERCHARGING THE FLATHEAD

This chapter discusses supercharging in general with specific application to the flathead.

Figure 4-1: Roadrunner Single Belt Kit installed on a 1949-1953 Flathead.

THE QUEST FOR MORE

We saw in the previous chapter that we can increase our flathead's horsepower by 35-45% on top of whatever else we've already done by supercharging! Of all the options available for the street, I believe that "blower is better." This option has been a proven, reliable horsepower adder for more than 60 years. Many types of crankshaft driven superchargers/blowers have been used including centrifugal, axial flow, vane, rotary screw (Lysholm), and Roots-type units such as those used on GM two stroke diesels for years. Exhaust driven turbochargers have also been popular.

Figure 4-2 illustrates the basic operating mechanisms for the most common crankshaft-driven supercharger types. A good discussion of street

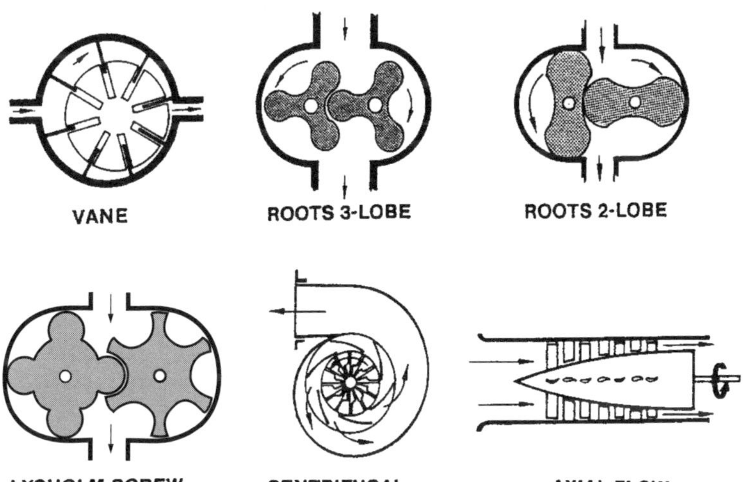

Figure 4-2: The most common types of superchargers used in automotive applications.

superchargers and supercharging in general can be found in Flathead Reading 6 in Appendix F.

In the U.S., Auburn, Cord, Duesenberg and Graham produced factory stock centrifugal type supercharged automobiles in the 1930's. The supercharged Cord had a 288 cubic inch flathead V-8, rated at 170 hp in 1937! The 1954 and 1955 Kaiser Manhattans incorporated the McCulloch centrifugal type supercharger to raise horsepower on their 226 cubic inch 6-cylinder from 118 to 140 hp. The 1957 Golden Hawk offered a McCulloch supercharged 289 cubic inch, 285 hp V-8 option which continued in various Studebakers including the Avanti model through 1964. When Ford found out they were behind the Chevrolet power curve in 1957, they produced the 300 horsepower McCulloch supercharged Y-block engines as an option for the Thunderbird and other Fords. Ford did it again in 1989 with the Thunderbird SC coupe which continued through 1997 with the Eaton Roots-type superchargers on the 3.8 litre V-6.

Turbochargers have also been factory options on several American and imported cars. The turbocharged Buick Grand Nationals are probably the most notable American examples. Turbochargers are primarily of the centrifugal type and are driven by exhaust gas turbines rather than by direct mechanical connection (e.g. belt, gear) to the crankshaft. Although turbochargers are extremely effective peak horsepower enhancers, they are much more difficult to engineer to provide good street performance over widely varying engine

Figure 4-3: S.CO.T. Supercharger and engine for Dan Kilcup's 1940 Coupe (1955). Note the stock-type heads.

speeds and loads than direct driven superchargers. We will only consider direct driven superchargers in this book with heavy emphasis on the Roots-type because of their widespread availability, moderate cost and good efficiency over broad engine operating conditions encountered on the street.

Hot rodders have been supercharging since the 1930's, but supercharged rods were not very common until the 60's. Today, a large aftermarket of most types of superchargers exists for the current crop of autos and many of the systems are emissions-legal even in California, which has the toughest standards. Beside being effective horsepower adders, reliable and environmentally friendly, supercharged engines are generally as fuel efficient as any normally-aspirated engine of equivalent performance.

So why doesn't everybody have a supercharger? One reason is expense. Where additional cubic inches are easy to get, bigger is generally the cheaper way for more power. There are many "big" engines out there today that produce adequate power without supercharging. For racing, there are other, cheaper options such as methanol and nitromethane fuels, "chemical superchargers," which are mechanically simple and readily available. Nitrous oxide is another proven short burst horsepower adder. Indeed, for racing, combinations of supercharging and "hot" fuels, and/or nitrous oxide are used for all-out applications.

Perhaps another reason supercharging is not more common is ignorance. Putting together a total system requires sound engineering and the use of best practices gleaned from expensive experience. The potential gains in power output are huge but the risk involved in going beyond the capabilities of the engine are also substantial. The same is certainly true in the nitro and nitrous camps, i.e. you can get too much of these good things if you don't know what you are doing. The rest of this chapter will attempt to help you maximize your flathead's output and minimize the risks.

EFFECTIVE CYCLE PRESSURE - THE SUBSTITUTE FOR CUBIC INCHES

As we hinted in Chapter 3, there is a substitute for cubic inches to increase torque. Just as we noted for cubic inches of engine displacement, increasing the mean effective cycle pressure (MEP) of the engine produces proportional gains in power output. Thus if we increase the pressure 50%, we increase the output power 50%, if the temperature and all other things are equal. The MEP can be defined as the average pressure on the piston available to do work. Pressure times area of the piston equals the force on the piston which is converted to torque on the crankshaft, which in turn becomes power when the crank rotates at a given speed. During a normal engine power cycle, air is inducted, compressed, expanded and exhausted. We can increase the mean effective cycle pressure by increasing the induction pressure and/or by increasing the compression ratio. Increasing the compression ratio above certain limits adversely affects flathead breathing as discussed in Chapter 3. In addition, increasing the compression ratio on supercharged engines can rapidly increase the peak or **maximum** cylinder pressure beyond safe limits.

The maximum pressure "spike" (short duration increase) resulting from **excessive** compression ratios on supercharged engines causes engine damage, blown head gaskets, detonation, etc. without a corresponding increase in the **mean** effective pressure or power output. Thus, in this chapter we will concentrate on increasing the induction pressure as the preferred path to increased power. Compression ratios of about 7.5:1 provide a reasonable compromise between power, reliability and fuel economy for **blown** flatheads.

A supercharger is an air compressor which raises the induction pressure (and temperature). The increased induction pressure and resulting combustion pressure is available for expansion and producing output power. The limitations in raising the induction pressure are the efficiency of the supercharger (goes down with increasing compression, heating the air more) and the structural limits of the engine. Induction pressure is also limited by detonation and pre-ignition and these are in turn a function of compression ratio, induction temperature, fuel octane, ignition timing, and other factors. The trick is to maximize MEP by maximizing induction pressure while staying within the practical constraints of the engine and available fuel characteristics.

PRE-IGNITION AND DETONATION

Pre-ignition and detonation are the most common and disastrous problems for a blown engine. (Pun intended!) Pre-ignition is an early and usually unintended ignition of the fuel-air mixture. The result of firing the mixture too early in the power cycle is wasted energy and potentially extreme structural loads on the piston and other rotating parts. Pre-ignition is the result of overheated spots in the combustion chamber, such as the spark plug electrode, other sharp edges, the exhaust valve, or areas near any air or steam pockets in the head cooling passages, etc.

Detonation is an extremely rapid, uncontrolled burning (actually an explosion) of the fuel air-mixture which may occur at multiple sites in the combusion chamber and either before or after normal ignition. The results and causes are often identical to those of pre-ignition. Detonation can be caused by many factors such as early (too much ignition advance before top dead center) ignition timing, excessive temperatures and pressures, insufficient fuel octane, "hot" spots in the combustion chamber, and others. Excessive temperature and pressures in turn result from over-boosting (over-supercharging), too high a compression ratio, an inefficient supercharger, and high intake air temperatures (e.g. under the hood on a hot day). Actual fuel octane can be degraded by excessive oil burning (worn rings, valve seals, blower oil seals, etc.).

The symptoms for both pre-ignition and detonation include low power, overheating, "running on" after the key is turned off, audible "pinging" or "knocking", blown (that word again) head gaskets or total destruction of the engine. Pinging and lowered power are usually the first symptoms of pre-ignition or detonation but may not be obvious at high engine speeds until it's too late!

Pioneers in supercharging learned the limits by destroying parts. We can do better.

PREVENTING PRE-IGNITION AND DETONATION

Ignition timing, air-fuel ratio, engine compression ratio, supercharger boost pressure and fuel octane are major variables affecting pre-ignition and detonation which we can control. Timing and air-fuel ratio will be covered under tuning in Chapter 5. For now we will assume that timing and the air-fuel ratio are optimized. We are interested in engine compression ratio because it is a very important variable affecting peak cycle temperatures and pressures. Induction pressure (atmospheric plus supercharger boost pressure) and temperature are likewise important for the same reason.

As a very general guide, for sea level, low humidity, 77 degree F blower inlet air, the minimum octane required for gasoline to prevent pre-ignition and detonation in our flathead is shown in Figure 4-4. Compression Ratio (CR) is the actual or estimated (see Appendix B) volumetric compression ratio of the engine. Pump gasoline octane is the average of Research (R) and Motor (M) octane, i.e. (R+M)/2.

Figure 4-4: Octane Requirements and Horsepower Vs Supercharger Boost and Engine Compression Ratio

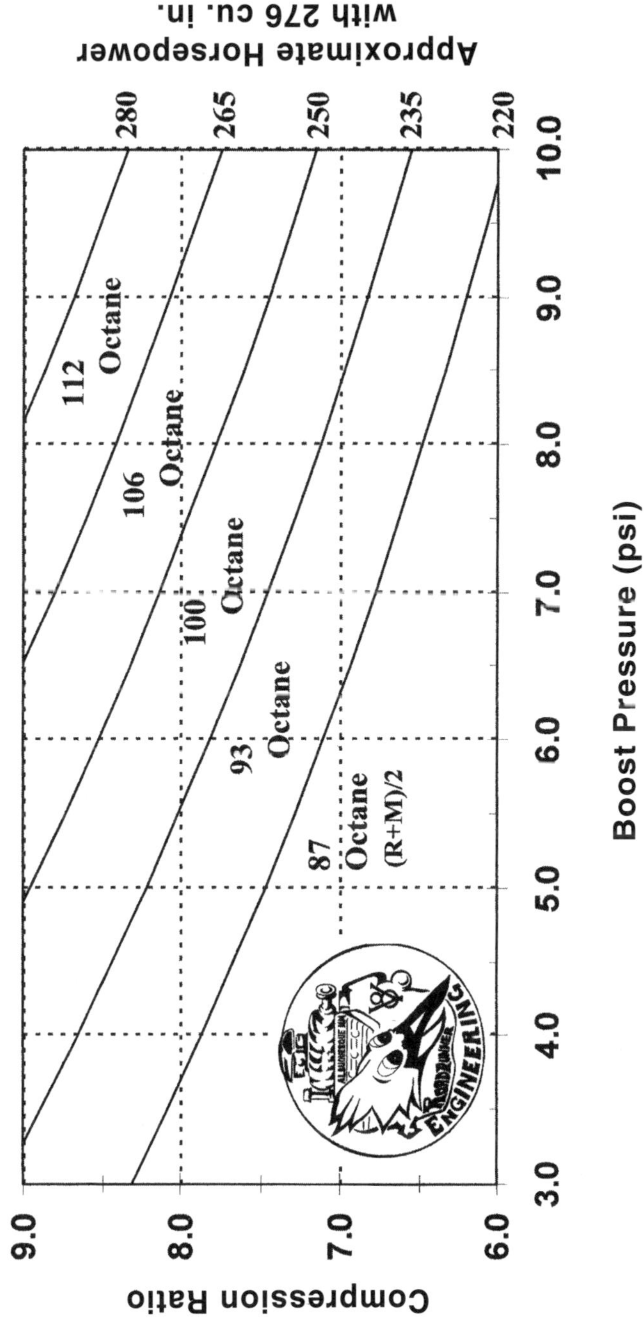

If humidity is greater than zero, the octane required is reduced. If atmospheric pressure is lower (altitude higher), it also reduces the octane required. Albuquerque and Denver are about 5280 feet above sea level and atmospheric pressure is about 12.4 lb/in^2 or about 15% lower than sea level. However, due to the higher compression heating to get the boost pressures in Figure 4-4, the octane required is about the same as for sea level. If blower inlet air temperature is greater than 77 degrees F (it could be 177 degrees F under the hood in the summer), we must reduce the maximum boost levels or increase the octane of the fuel.

BOOST & COMPRESSION RATIO vs POWER

Note the horsepower figures associated with the octane lines in Figure 4-4. The primary determining factors for octane requirements are the pressure and temperature in the combusion chamber. These same cylinder pressures and temperatures, along with the engine displacement are in turn primary factors in determining the power output of the engine. Thus if we know the engine displacement, we can associate the octane lines with an estimate of the power output of the engine. The horsepower figures shown in Figure 4-4 are all rough approximations for a midperformance, 276 cu. in. (e.g. 3-5/16" bore x 4" stroke) displacement engine. Horsepower estimates for bigger or smaller displacement engines can be made by taking the ratio of the actual displacement over 276 and multiplying this ratio by the numbers on the curves for the 276 engine. Again, I emphasize that these estimates are only approximate, but they can be useful for planning our engine buildup.

We note that for the street, about 6-7 lb/in^2 boost with an 8:1 compression ratio engine requires 100 octane gasoline at sea level. 100 octane aviation or race gas costs about 50 -100% more than 93 octane premium automobile gas. We can also use gas additives or water injection to get the required equivalent octane by trial and error. However, limiting ourselves to 6 lbs/in^2 boost and a 7.5:1 CR allows us to use premium pump gas and still get about a 40% horsepower boost over and above whatever the engine was producing normally aspirated.

Although not important for a street engine, alcohol can tolerate high boost pressures and/or high compression ratios even though its octane rating is only about 100 or so, because it has a much greater cooling effect than gasoline.

For racing, where fuel economy and fuel cost are not important considerations, maximum power with minimum risk can be obtained with very high boosts and very low compression ratios. Typical values for alcohol-nitromethane fuel burning engines might be 20 lbs/in^2 boost with 5.5:1 compression ratios! The high boosts and low compression ratios for these engines promote high volumetric efficiency and high MEP while minimizing the destructive maximum pressure "spikes" associated with higher compression ratios discussed earlier.

BOOST AND BLOWER SIZE RECOMMENDATIONS

Based on the considerations in the previous section, I recommend 6 lbs/in^2 boost with a 7.5:1 CR as a practical combination for everyday driving at sea level. This allows the use of readily available unleaded premium fuel of 93 octane and the engine can be essentially stock if everything is up to specifications. For weekend drag strip duty, higher boosts up to 10 lbs/in^2 or more with a 7.5:1 CR delivers excellent performance, but requires racing gas, forged pistons and at least minimal center main bearing support (e.g. a "strap").

The boost, of course, can be tailored by the size of the blower, and the blower and crank pulley combination. Thus, if the rest of the engine is built to take the high boost, you can run more aggressive pulley combinations on the weekend. It should be noted that boost pressure is not only a function of atmospheric conditions and the blower capacity but also the breathing capacity of the engine. A "free breathing" engine will "inhale" and run better while indicating a lower boost pressure in the manifold than a more restricted engine, all other things being equal. It is not unusual for the boost pressure to increase by a factor of two or so at wide open throttle with a positive displacement supercharger going from 2500 to 5000 rpm. A lower increase indicates a better breathing engine, all other things being equal. Breathing is influenced by the air cleaner, carburetion, blower characteristics, intake and exhaust manifold design, cam and valve train, and head and block porting parameters. Although boosting the flathead with a supercharger helps overcome it's basic breathing restrictions, better breathing means better performance either blown or unblown.

Roots-type blower capacities of about half to two-thirds the engine displacement per revolution work very well on the flathead and are plenty big enough for street and weekend drag strip duty. Smaller units require higher rpms to increase capacity with a corresponding decrease in blower life. Bigger units tend to be inefficient due to rotor leakage losses at lower rpms with high boost, but may be favored for sustained high speed driving. An oversized supercharger which is underdriven sufficiently to prevent over boosting at high speeds may actually "get in the way" at lower engine speeds! The choice of the blower and crank pulley sizes are of course dictated by our performance goals, atmospheric pressure, the size (displacement) of the engine and its breathing characteristics, the size of the blower and its characteristics and other factors such as fuel cost, etc. The blower kit suppliers, including my company, will help here. Table 4-1 lists characteristics of various Roots-type blowers.

Blower/Supercharger Make & Model	Capacity (cu. in./rev)	Rotor Design
GMC 2-71	136	3 lobe twisted
GMC 3-71	210	" " "
GMC 4-71	280	" " "
GMC 6-71	420	" " "
GMC 4-53	139	2 lobe straight
GMC 6V-53	208	" " "
GMC 8V-53	334	" " "
Italmeccanica/S.CO.T. (8.63" long rotors)	172	2 lobe straight
Camden 7"	112	2 lobe straight
Camden 9"	144	w/Teflon tips
B&M/Holley 144	144	2 lobe straight
B&M/Holley 174	174	w/Teflon tips
B&M/Holley 250	250	
Eaton M112	112	3 lobe twisted
Weiand Pro-Street 142	142	2 lobe straight
Weiand Pro-Street 177	177	" " "
Weiand Pro-Street 256	256	" " "

Table 4-1: Air moving capacity (displacement/revolution) and rotor design of selected Roots-type blowers/superchargers.

As an example of sizing, in combination #9 in Table 3-1, a 276 in^3 engine with a 144 in^3 blower would require a blower pulley of 4 inch pitch diameter paired with a 6 inch pitch diameter crankshaft drive pulley, (6/4 or 1.50 ratio or 50% overdrive), to produce 260 hp with 6.6 lbs/in^2 boost in a free breathing engine at 5000 rpm at sea level.

Figure 4-5: Roadrunner Nostalgia Kit installed on a 1938-1948 Flathead.

5

BUILDING A COMPLETE PACKAGE

This chapter discusses the key ingredients for a high output supercharged flathead capable of 200-300 horsepower. Choosing the right block and the supercharger parameters are covered in Chapters 2 and 4. The other subsystems are covered below.

BLOCKS

See Chapter 2 for basic preparation and checking of the block.

For supercharged engines, it is neither desirable nor cost effective to bore, port, relieve or otherwise modify the block excessively, such that structural integrity and reliability is lost. Boring the cylinders in excess of 0.125" is not recommended. Likewise, restraint should be used in porting and relieving to assure that the block wall structure is sufficient for the much higher cylinder pressures resulting from supercharging. See section in this chapter on Porting and Relieving.

PISTONS AND RODS

Boosts up to about 6 lbs/in^2 are compatible with original equipment manufacturer (OEM) quality pistons and rods. Cast pistons should be adequate for the above boost levels. Low expansion silicon aluminum alloys provide a quiet engine and promote good oil control with either 3- or 4-ring pistons. The so-called "hypereutectic" type pistons provide an extra strength margin over OEM casting alloys.

Cast iron, "moly" or chrome plated rings are all fine as long as ring manufacturer's recommendations are followed for cylinder bore finish. Care should be taken to assure ring end gaps of 0.012" to 0.016". "Spiroloc" wrist pin retainers are recommended. Higher boost levels (over about 6 lbs/in^2) require forged aluminum pistons for strength. The forged pistons require significantly more clearance in the bores (0.004"-0.007" typical) versus the low expansion cast alloys (0.0015"-0.003" typical). Piston-to-head clearance should be .040" typical in the area farthest from the valves on all engines.

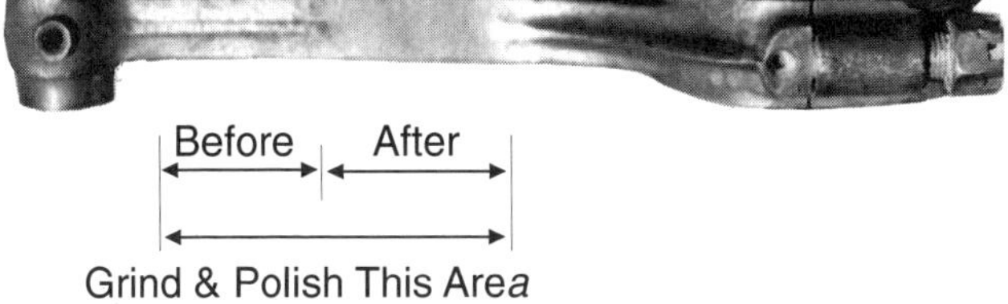

Figure 5-1: Stock Flathead Rod Showing Polished and Unpolished Beam Side.

Excessive piston-to-head clearances (e.g. 0.1" or more) in the area over the piston, furthest from the valves promote detonation. See Flathead Reading 21 for an excellent discussion of flathead combustion chamber design.

Stock rods do not look very stout but are adequate for less than all-out applications. The rods should be sanded and polished on the flat sides of the "I-beam" to remove stress concentrations. See Figure 5-1. Shot peening and a Magnaflux check for cracks are also a necessary part of reconditioning.

High strength rods are available from several of the suppliers in Appendix G.

CRANKSHAFT & MAIN BEARINGS

Stock crankshafts are high quality steel castings and again are adequate for less than all-out applications. Some people have asserted that some flathead cranks are forged, but my research of factory information has revealed that only the 1932 cranks and the big Lincoln V-8 flathead cranks were forged. The crank should be thoroughly cleaned and magnafluxed, of course. The crankshaft should be ground no smaller than 0.030" undersize. Machined billet high strength steel cranks are available on the aftermarket in strokes up to 4-1/2" but require special rods and/or pistons for the longer strokes.

The flathead has only three main bearings

versus five for most modern V-8's. This means high loads especially in the center main which carries twice the load of the end bearings. Although the stock main caps are robust, extra support or an aftermarket main bearing cap is recommended for boost pressures much over 6 lbs/in^2.

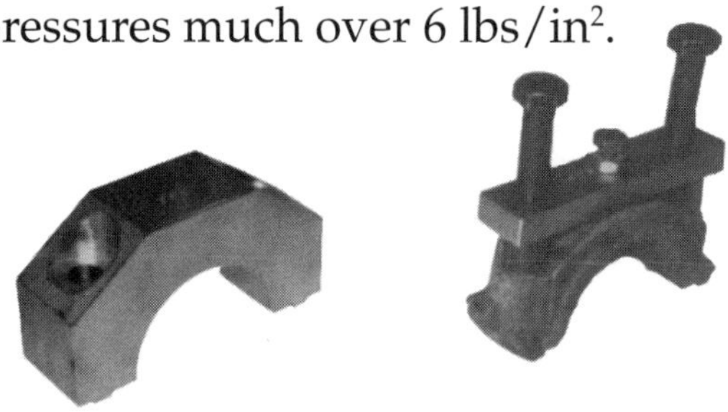

Figure 5-2: Aftermarket High Strength Main Bearing Cap (left) and Stock Cap with Aftermarket "Strap" (right).

Custom main caps are even available with 4 bolts if desired. I have had good results with a simple 1/2 inch thick "strap" support available from Offenhauser dealers. These are inexpensive and provide cheap insurance. See Figure 5-2.

"Clevite" or "Michigan" 77 rod and main bearings are recommended with 0.0015"-0.0025" clearance between the crankshaft and non-floating rod bearings and between the crankshaft and the main bearings.

CAM AND VALVE TRAIN

The camshaft is a very important part of the power equation and the flathead is a particularly receptive candidate for improvement in this area. There are several grinders for flathead cams and some grinders will grind custom profiles or reproduce grinds produced by grinders in the past who are no longer in business such as Potvin and Winfield. One grinder who has produced flathead cams continuously since the 1940's is Iskenderian Cams. I have had excellent success with their products. The Iskenderian 400 Jr. grind with single valve springs (P/N 185G) was personally recommended to me by Ed Iskenderian for blown street engines and has given excellent results in my own car. The 400 Jr. is considered a "full race" cam with a 0.400" valve lift and approximately 280 degree duration @ 0.016" tappet clearance (244 degree @ 0.050" lift). The lobe center angle is 111° for this cam. This cam has a definite lope at idle but is still quite streetable with good vacuum even at low speeds. The approximately 11 degree intake/exhaust overlap @ 0.050" lift for this cam is acceptable and reduces the probability of potentially blower damaging backfires. Isky's 77-B or Max 1 cams are recommended for those who want something tamer than the 400 Jr. The stock factory cams from 1952 and 1953 are not a bad choice when available. They have 0.037" more lift and 40° more duration than earlier stock cams.

Perhaps one of the most powerful racing cams for the flathead was Isky's 404, but these cams require keyed and "radiused" lifters resembling roller cam lifters which are not commonly available. The Winfield SU-1A and the Potvin "Eliminator," are also popular cams for all-out flatheads.

Cam	Valve lift	Duration @0.050" int/exh	Overlap @ 0.050" lift	Lobe separation angle	Lash (cold)
Stock EAB or EAC	.314/ .310	231/221	-1°	114°	.014/ .018
Isky 77-B	.325	220	-1°	111°	.014
Isky Max 1	.364	226	2°	111°	.014
Isky 400 Jr.	.400	244	11°	111°	.016

Table 5-1: Cam Comparison Chart.

Single heavy duty valve springs such as the old Lincoln Zephyr and those available from Isky and others are adequate for most applications up to 5200 rpm even when used with the heavy, solid-type Johnson adjustable lifters, also available from Isky and others. Dual valve springs are only necessary for higher rpm applications. One advantage of a flathead is that the engine does not self destruct if the valves float. Valve float can cause dangerous backfires, however.

The stock Ford 1.5" diameter valves are very good items but special high flow stainless steel valves are available from Manley and others. These special valves are available specifically to flathead specifications but those made for small block Chevy exhaust valves work great. The Chevy type valves are available in 1.5", 1.6", and 1.72" diameters and they are about 0.10" longer than the Ford valves. This extra length works well with a high lift cam since the base circle of these cams is approximately 0.10" smaller than stock anyway and the difference is well within the adjustment range of the lifters. The benefit from larger than stock valves is primarily at low to mid lifts and generally requires additional head clearance around and over the valve (0.04" minimum). Assuming that proper head clearance is provided, the 1.6" diameter valves are recommended for both intake and exhaust on supercharged street flatheads. See Chapter 3 and Appendix A for flow benefits.

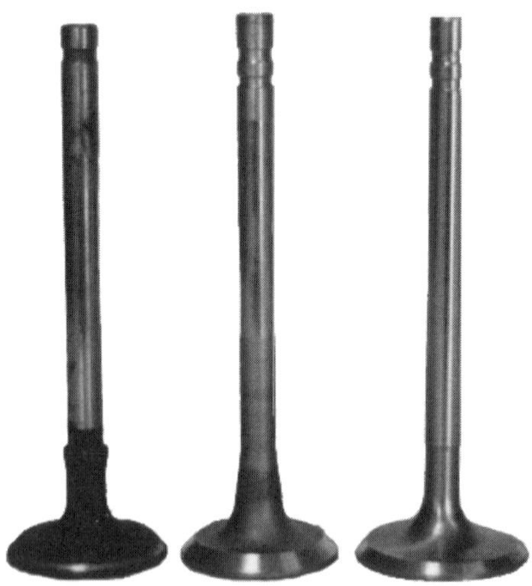

Figure 5-3:
Stock Ford Valve (left); Chevrolet Type Standard Aftermarket Replacement Valve (center); High Flow Valve (right).

PORTING AND RELIEVING

Port matching of the intake manifold and block is absolutely essential to good performance. A study of the flow test information in Chapter 3 and Appendix A illustrates the gains to be made by various steps and degrees of porting and relieving. From all the data, the following recommendations are offered. Refer back to Figure 3-3.

1. Smoothing the "short side" edge ("B" in Figure 3-3) of the intake port is an easy, high payoff modification.
2. Relieving the block ("C" in Figure 3-3) to the head gasket boundaries is a significant improvement, particularly if high compression aluminum heads are used. Maximum relief depth should be limited to 0.100" for street flatheads. Some factory blocks had 0.80" deep reliefs which are adequate. General smoothing even with no relief is also adequate with stock or equivalent low compression heads.
3. Rounding the inside edge of the end exhaust ports where they exit the block is also an easy, high payoff modification. (Otherwise the end ports flow only 80% of the center ports.) See Chapter 3 and Appendix A.
4. Extensive porting should be done carefully and is best done by an experienced professional.

HEADS AND HEAD GASKET

Appendix B lists the combustion chamber free volume for several stock cast iron heads and some commonly available traditional high compression aluminum heads. The free volume numbers can be used to plan or estimate the compression ratio for your engine as described in Chapter 3 and Appendix B.

Some of the trade-offs between aluminum and cast iron heads are as follows. The traditional aluminum heads provide an attractive focal point for your flathead, i.e. they really look nice. They can provide an increase in low rpm performance particularly for normally-aspirated engines. They are lighter by about 10 lbs/set. They may also be less prone to hot spots due to aluminum's high conductivity. Many of the aftermarket aluminum heads will accept larger than stock valves and high lift cams without additional machine work. The downside to aluminum heads include cost (currently about $400/set), difficulty in sealing with high combustion pressures (e.g. blower or fuel engines) and a proneness to corrosion and electrolysis.

Iron heads by comparison are almost "free" and commonly available. They breathe or flow very well, are easier to seal because of their stiffness (3 times that of aluminum), and they match the thermal expansion of the block. Sealing is also enhanced by the higher head bolt torques recommended for the iron versus aluminum heads (usually 70 ft-lbs versus 55 ft-lbs, respectively). The iron heads also keep the heat in the combustion chamber better than aluminum due to their lower conductivity.

The earlier iron heads may require some machining to provide clearance and breathing room for big valves and high lift cams. An unmilled stock head will provide the requisite 0.040" overhead clearance for stock 1.5" diameter valves up to a lift of about 0.375". Above these numbers may require some minor machining. Cams with 0.400"+ lift will also require spacers under the spark plugs or shorter plugs (e.g. 3/8" reach versus 7/16" reach) to prevent the valves from closing the gap. With the shorter plugs, the exposed plug threads inside the combustion chamber should be removed by chamfering. Overall, the stock-type iron heads are recommended for performance and cost for supercharged engines.

The "best" head gaskets seem to be as follows in order of preference:

1. Stock type composite gaskets with steel fire ring around the combustion chamber. Good for bores up to 3-3/16" + 0.060".
2. Felpro (P/N 1055/1056 for 1949-1953) or Victor (P/N 3036 for 1939-1948) copper sandwich type "big bore" gaskets for stock up to 3-3/8" bores.
3. Composition gaskets similar to some high performance intake manifold gaskets. These have a shiny, pencil-lead color surface and a steel screen mesh reinforcement in the middle and no fire ring. Low stress applications only.

High performance engines utilize much higher cylinder pressures which require a good head-to-block seal. With any head or gasket combination, the retaining bolts or nuts should be torqued and re-torqued a minimum of three times before full throttle is attempted with a normally aspirated engine and five times with a supercharged engine. The engine should be allowed to warm up to operating temperature and then cool down between torque cycles.

CARBURETION AND FUEL SYSTEMS

A Stromberg "97" carburetor or Holley "94" 2-barrel type carburetor will flow approximately 155 cubic feet per minute (cfm) at 1.5 inches of mercury pressure differential. Two of either of these carburetors or a Holley 390 cfm (rated at 1.5 inches of mercury) 4-barrel is adequate for mild street applications. Three or even four 2-barrel carburetors or a 600 cfm 4-barrel will work for higher performance applications. To use the three 2-barrel carburetors or the larger 4-barrel on the street, a progressive type linkage or vacuum-operated secondaries, respectively, are recommended.

A rich mixture enhances power and inhibits detonation at the expense of fuel economy. Generally, one to four sizes larger than stock on the main jets gives good results and prevents lean out damage.

In a 4-barrel carburetor, the secondaries can be jetted significantly richer than the primaries with little effect on mileage during normal driving. The power valve jetting and pressure activation level provide additional tuning aids to get optimum mixtures under full throttle.

A high capacity electric fuel pump is necessary for all high performance applications. A good fuel filter and pressure regulator are highly recommended for all applications. The old 2-barrels can only tolerate 2-3 lbs/in^2 fuel pressure for reliable float operation while 5-7 lbs/in^2 is fine for Holley 4-barrel carburetors. A 3/8" inside diameter, or larger, fuel line is recommended between the tank and the regulator. If an external regulator is used, it should be mounted close to the carburetor(s).

IGNITION AND TIMING TIPS

The 1938-1948 flathead used a front mounted, direct camshaft driven distributor with a centrifugal mechanical advance, and a vacuum brake (not retard). The vacuum brake slows or restricts the movement of the centrifugal advance mechanism under low vacuum (e.g. wide open throttle at low speed) engine conditions. Since even the worst gas now (e.g. 87 octane) is better than the best gas (e.g. 78 octane) in 1948, the brake operation is not needed and the adjustment is fully backed off. The stock distributor works fine but needs a higher voltage coil for use with a blower. Even the stock points are adequate to engine speeds of 5200 rpm or so.

The 1949-1953 distributor is mounted up high on the right front of the engine and is gear driven from the cam, similar to most modern distributor systems. These "Load-O-Matic" distributors have a vacuum-only operated advance mechanism and give poor results with multiple carburetion and/or blower setups. All is not lost, however, as Mallory, MSD and others make centrifugally operated advance distributors for the 1949 and later applications. These distributors are available with dual mechanical points, or electronic or optical switching, and are compatible with high output coils, which are a must.

Early (pre 1949) type distributors can be run on 1949 and later engines and vice-versa simply by changing the front engine covers, the camshaft (to account for the opposite thrust load on the cam for the two different systems) and the crankshaft timing gear.

Typical stock ignition tuning specifications for all '48 and earlier Ford flatheads are 2 degrees initial and 13 degrees total advance at the distributor which is equivalent to 4 degrees initial and 26 degrees total (at 3450 rpm) at the crankshaft. Most of the early distributors were timed off the engine (i.e. on the bench), which is why the tune-up specs are usually given in distributor degrees for the early models. The later flatheads called for 2 degrees intial and 18 degrees total (@ 4000rpm) at the crankshaft.

There is an optimum ignition timing for every combination of engine, fuel and operating condition. The Ford flathead is a high turbulence, "fast burn" engine and does not require as much advance as most overhead valve engines. In addition, as cylinder pressure goes up so does flame speed and the engine requires less advance for peak power. Blown, gasoline fueled flatheads seem to run well with only 0 to 4 degrees initial advance and only 16 to 20 degrees total advance at the crankshaft. Running on the low side of this range does not seem to hurt power and gives added insurance against hard-to-detect high speed detonation. Unfortunately, it is difficult to determine optimum settings without testing, so be conservative!

For supercharged engines, consideration should be given to retarding the ignition timing automatically when the engine is under boost (e.g. one degree/psi boost). This allows the use of lower octane fuel and/or more ignition advance for normal driving (i.e. not wide open throttle). Several aftermarket suppliers can provide the boost retard feature or even fully programmable ignition modules compatible with any ignition system.

Some alcohol flathead racers recommend 4 degrees additional advance over gas numbers for optimum performance and often use locked (no automatic advance) distributors or magnetos. Flatheads with "high flow" type heads with large flow passages and pop up pistons may require different timing specs for peak performance.

EXHAUST

A free flowing exhaust is a definite performance enhancer for stock flatheads and is a must for peak performing blower engines. The stock flathead headers are candidates for the most restrictive items of all time. Tubular or cast iron headers are available from several of the suppliers. See Appendix G. Center dump ("Belond" or "W" type) tubular headers are an excellent choice for high performance street flatheads or an all-out racer. These headers, however, will not fit 1948 or earlier cars with stock wishbones in the original location.

Figure 5-4: Center-dump headers, pictured above, are recommended where space allows.

Center exhaust port dividers are recommended by some builders to enhance low rpm torque and inhibit exhaust-intake interaction on the center cylinders, but I have not seen any dynamometer data to support the benefits. There is no measurable effect on exhaust flow with these dividers. See Appendix A.

MISCELLANEOUS TIPS

OVER REVVING

The 59A and 8BA flatheads will never be high revving engines despite our best work. The long stroke and long skinny rods (not to mention the three main bearings) limit street flatheads with mostly stock parts to 5200 rpm (4" stroke). This is ok because breathing has already drastically fallen off by this point anyway, with peak torque for street engines having occurred at 3500 rpm or less. Supercharged flatheads have a very broad torque curve which does not fall off as quickly as one for a normally aspirated engine. This encourages over-revving – – don't do it! Figure 5-5 shows the piston acceleration loads versus rpm for a typical 8BA engine. Note that these loads go up as the rpm squared. On the firing or compression stroke, the cylinder pressure helps cushion these loads but are no help on the exhaust stroke.

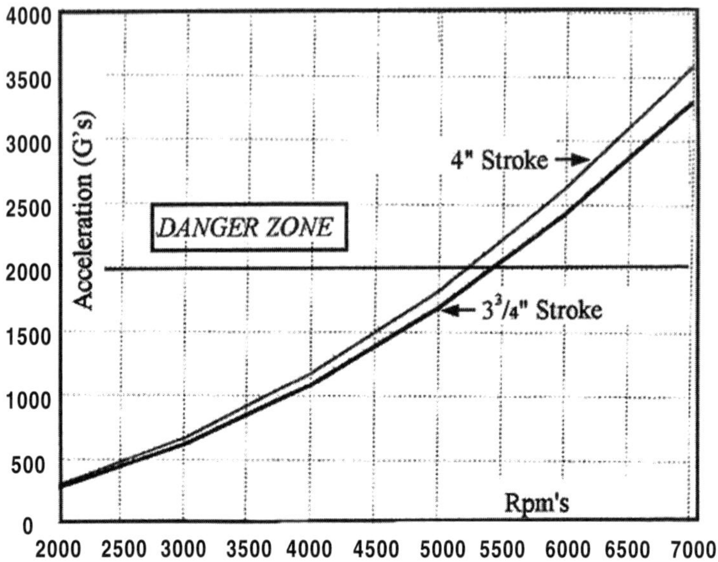

Figure 5-5:
Piston Acceleration (G's) Versus Engine Speed in RPM
Accelerations above 2000 G's
should be avoided for engine longevity.

DRIVE TRAIN

A healthy supercharged flathead street engine can produce 250 to 300 ft-lbs of torque. Most of the early drive trains were rated marginally for 200 ft-lbs. Stronger is better if you want to play hard. Adapters are readily available to use modern automatic and manual transmissions. Check the torque rating for candidate transmissions!

THE RESULT – *A "REAL" HOT ROD!*

Good things happen when we build a high performance supercharged flathead and we drop it into our 1932 Ford roadster. Going from combination #7 to #9 in Table 3-1 will increase our standing start quarter mile speed from 90 mph to 100 mph and reduce our elapsed time from 15 seconds to 13.4 seconds! Our 0-60 mph time drops from 6 seconds to 4.5 seconds!!! *WOW! What a great way to get rid of those pesky jelly bean cars in our rear view mirrors.*

Appendix A

FLOW TESTING WITH RESULTS

The inherent breathing limitations in the flathead design are obvious when we look at a cross-section such as Figure 3-3. Also as pointed out in Chapter 3, some of the enhancements to breathing hurt us by reducing the compression ratio. All this being noted, we can significantly improve the breathing of our flatheads.

Our approach to enhancing breathing must be balanced, i.e. we must work on each area of the airflow tract more or less equally. The original flathead design is balanced, i.e. there are no "golden bullets" or single improvements which will "unplug the pipe," although some improvements are easier to accomplish. As an example, bigger valves or relieving by themselves do not produce dramatic improvements in airflow. As we improve one area of the "pipe," another area becomes the limiting item which we can then improve and so on, until we have reduced flow resistance throughout the system. By working on each section of the flow passages in a balanced manner, we can improve flow 40% or more as shown in the tables and graphs that follow.

The curves and tables that follow are the result of over 100 modifications and flow tests performed by the author and Ron Meek of Flow Dynamics in Albuquerque. A flathead block and head assembly were connected to a SuperFlow SF-600 flow bench as shown in Figure A-1. All tests were run at 28 inches of water test pressure. Valves were prepared with a conventional 3 angle valve job with 45 degree seats. Center dump ("W" or "Belond" type) exhaust headers were used at the exhaust ports and the intake ports used a molded clay inlet prep.

INTAKE

Figure A-2 shows the flow and changes in flow observed as we increase the valve size from the stock 1.5" diameter to 1.6" and 1.72" diameters. Three of these curves include the effects of "rolling" or smoothing the short side radius ("B" in Figure 3-3) and undercutting or opening up the head area in back of the valves ("A" in Figure 3-3) with a 0.375" ball mill cutter. Both the rolling and undercutting are worthwhile, and are low risk modifications that can be performed by a non-professional engine builder. We note that the 1.72" valve actually flows less than stock even with a modified stock head because the bigger valve chokes off flow in back of the valves and there is insufficient room to open up this area in any head which uses the stock water passage location.

Figure A-3 shows the flow we observed with a "maximum street" porting and relieving job using modified stock heads. As shown in the figure, we can make dramatic changes by working on all the areas noted in Figure 3-3 (areas "A" through "E").

Figure A-1: Flathead & SuperFlow SF-600 Flow Bench Setup.

As we can see from Figures A-2 and A-3, there is no advantage in using intake valves larger than about 1.6" diameter in heads limited by stock type water passage location.

We note that combustion chamber **shape** as well as volume is important. The traditional aluminum high compression head with 63cc combustion chambers does not flow as well as the modified stock head because the smaller combustion chamber "transfer area" is more restrictive. Not all heads are the same however, whether stock or high compression. Thus, heads with similar combustion chamber volumes ***may*** have different flow characteristics. That is why we flow test!

As discussed in Chapter 3, flow is only part of the power equation, but for blown street flatheads, any of the stock heads are good choices for performance and durability.

The Offenhauser 425, or equivalent, is a reasonable choice when a traditional type aluminum high compression head is desired. These can be modified, within bounds, to improve flow. Relieving the block is recommended with this head for peak performance.

Figure A-2: Intake Flow Versus Lift with 8BA Head and Various Valve Sizes @ 28" H_2O Test Pressure

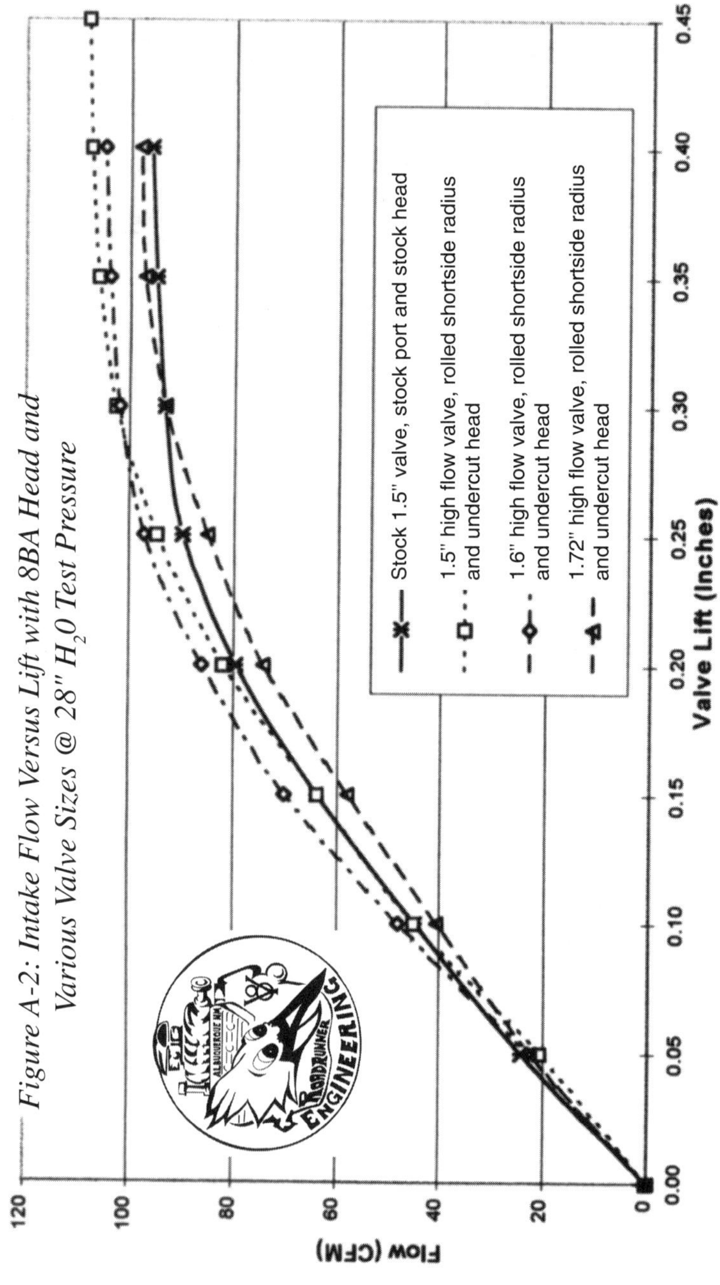

Valve Lift (Inches)	Intake Flow (CFM) with 8BA Head @ 28" $H_2$0 Test Pressure			
	Stock port & stock head	High flow valve, rolled shortside radius & undercut head		
	Stock 1.5" valve	1.5"	1.6"	1.72"
0.00	0	0	0	0
0.05	24.3	20.5	22.9	22.5
0.10	44.7	45	48	40.5
0.15	63.9	63.8	70.2	58
0.20	79.5	82	86	74.2
0.25	89.7	94.5	97	85
0.30	93.1	103	102	93
0.35	94.8	106	104	97
0.40	95.8	108	105	98

Tabular Data for Figure A-2

Figure A-3: Intake Flow Vs Lift with 8BA and Offenhauser 425 Aluminum Head with Various Valve Sizes @ 28" H_2O Test Pressure

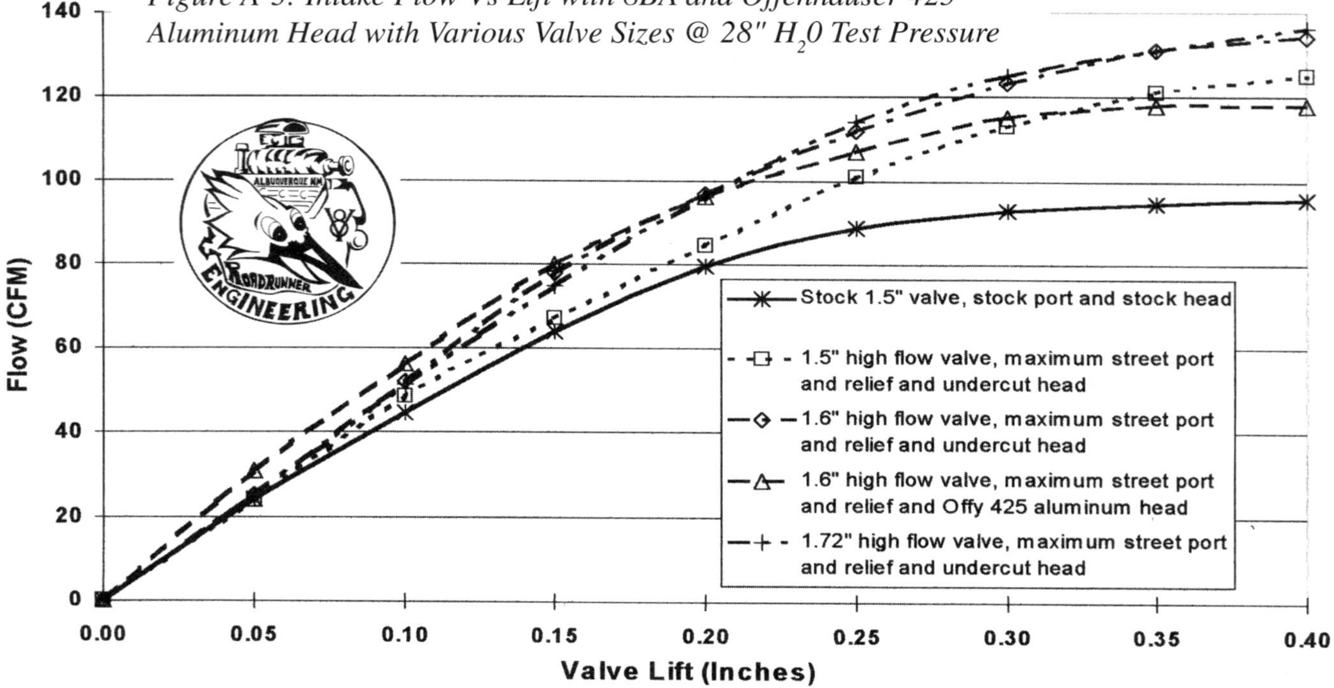

Valve Lift (Inches)	Intake Flow (CFM) @ 28" H_20 Test Pressure				
	Stock port & head	High flow valve, maximum street port and relief			
	Stock 1.5" valve	1.5" 8BA Head[a]	1.6" 8BA Head[a]	1.6" Offy 425 Head[b]	1.72" 8BA Head[a]
0.00	0	0	0	0	0
0.05	24.3	23.9	26.3	30.8	25.1
0.10	44.7	48.5	53	56	51.3
0.15	63.9	67	78.8	80	74.7
0.20	79.5	84.5	101	96	96.1
0.25	88.7	101	116	107	114
0.30	93.1	113	127	115	125
0.35	94.8	121	134	118	131
0.40	95.8	125	138	118	136

[a]Head undercut behind the valves as shown in Figure 3-3.

[b]Traditional high compression aluminum head with 63cc combustion chamber.

Tabular Data for Figure A-3

EXHAUST

The stock end exhausts flow about 80% of the middle exhausts, primarily due to a very unstreamlined dead end corner just inside the outlet ports at the extreme ends of the block. This is easily observed by looking in the end exhaust ports where they exit the block. Chamfering and smoothing of this area (outlet port) is easy and improves flow about 15% on the end ports.

Figure A-4 shows the effect on flow of different modifications of the end exhaust ports. Porting the end exhausts drastically improves flow in these ports. Note that the 1.6" diameter oversize valve does not have much effect on exhaust flow at peak lifts in the maximum street port and relief combination but does provide about a 10% benefit over the 1.5" valve with the combination where only the head is undercut and the outlet port is smoothed.

Figure A-5 shows the flow effects from modifications on the center exhaust port. Note that the installation of a cast iron airfoil shape divider in the center ports does not appreciably affect flow. The effect of this divider on performance must be determined on the dynamometer. Neither relieving nor the 63cc high compression aluminum head substantially affected exhaust flow because the major resistance to exhaust flow is downstream of the restricted cylinder-to-valve passage. Enlarging the center exhaust ports is limited in the passage area between the cylinders and thus limits the gains from porting on center exhaust flow.

Although the stock end exhaust ports initially flow less than the stock center exhaust ports, porting allows the end exhaust ports to equal or exceed the center exhaust ports.

Figure A-4: Flow Vs Lift for End Exhausts with 8BA Head with Center Dump Header @ 28" H_2O Test Pressure

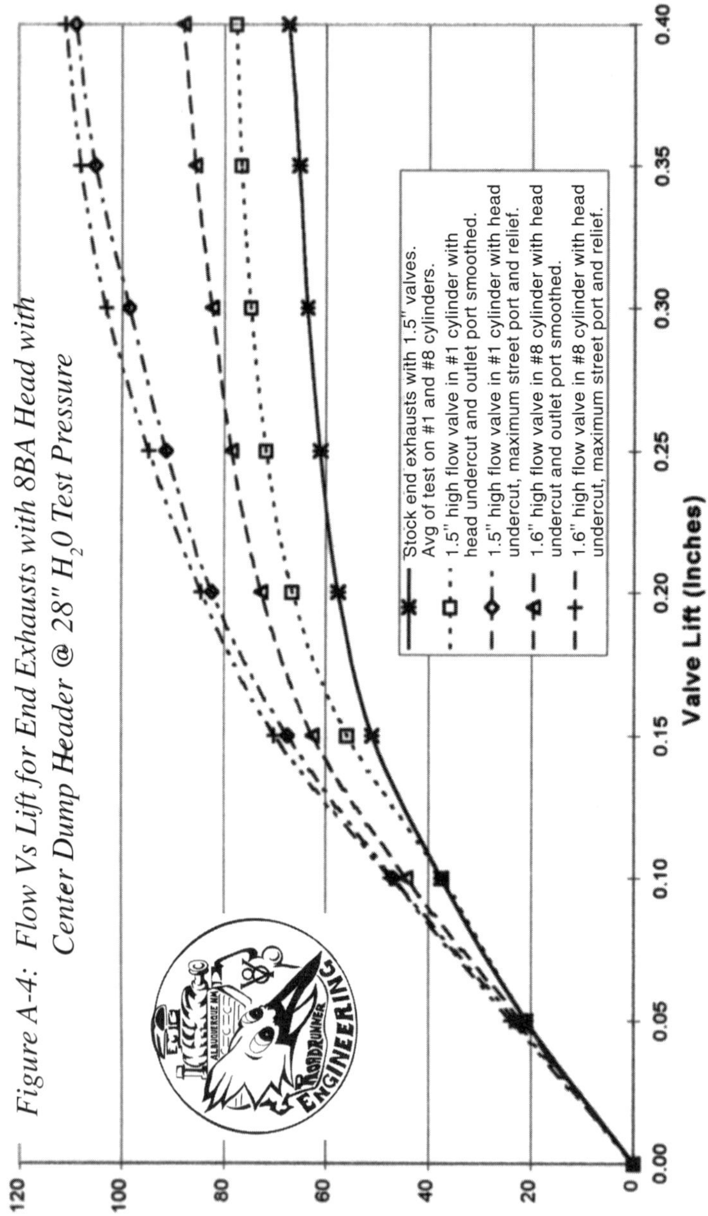

Valve Lift (Inches)	Exhaust Flow (CFM) with 8BA Head with Center Dump Header @ 28" H_2O Test Pressure				
	Stock end exhausts	Head undercut behind the valve and block outlet port smoothed		Head undercut behind the valve with maximum street port and relief	
	1.5" valves Avg on #1& #8 cylinder	1.5" high flow valve in #1 cylinder	1.6" high flow valve in #8 cylinder	1.5" high flow valve in #1 cylinder	1.6" high flow valve in #8 cylinder
0.00	0	0	0	0	0
0.05	21	21	22	23	24
0.10	37.5	37.3	44.3	46.9	47.4
0.15	51	55.7	62.4	67.5	70.1
0.20	57.5	66.5	72.7	82.3	84.5
0.25	61	69.7	78.5	91.3	94.7
0.30	63.5	74.7	82.4	98.5	103
0.35	65.2	76.6	85.7	105	108
0.40	67.3	77.6	88	109	111

Tabular Data for Figure A-4

Figure A-5: Flow Vs Lift for Center Exhausts with 8BA Head with Center Dump Header @ 28" H_2O Test Pressure

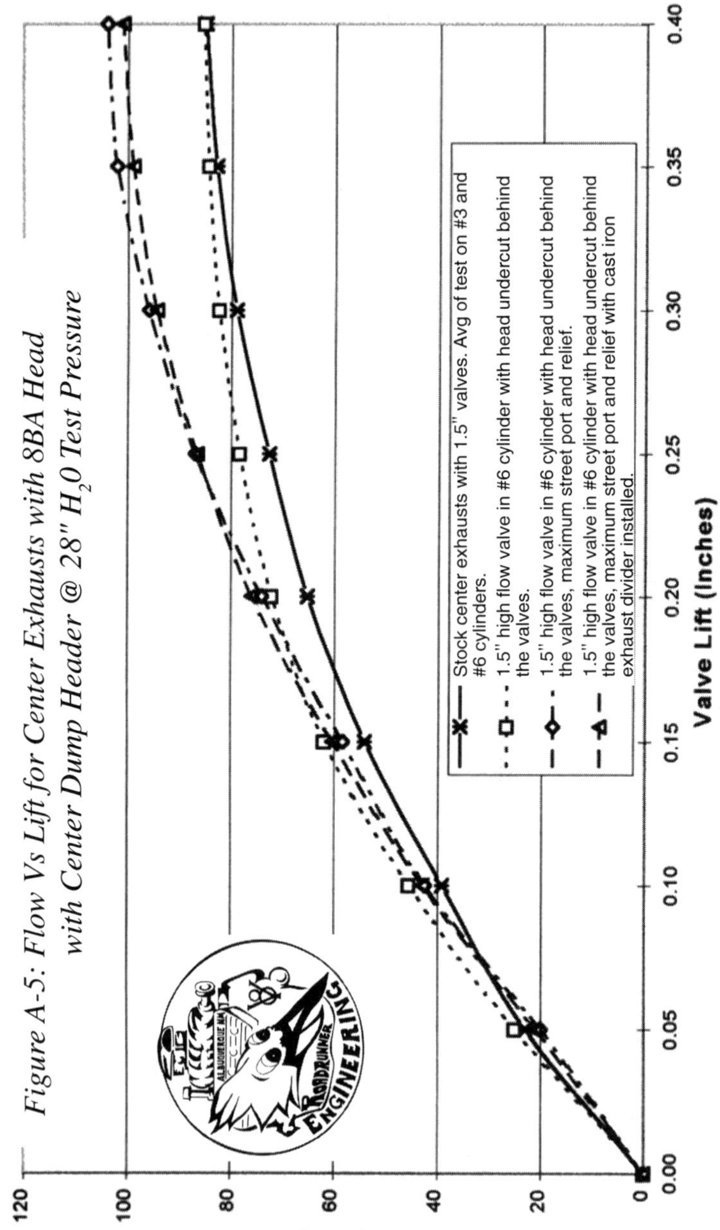

Valve Lift (Inches)	Exhaust Flow (CFM) with 8BA Head with Center Dump Header @ 28" H_2O Test Pressure			
	Stock 1.5" valve Avg on #3 & #6	1.5" High flow valve in #6 cylinder Head undercut behind the valves		
	Stock center exhausts	No porting or relieving	Fully ported and relieved	Fully ported and relieved with exhaust divider
0.00	0	0	0	0
0.05	23	25	20	21
0.10	39	45.4	42.3	42.8
0.15	54	61.8	58.3	60.3
0.20	65	72.1	74	76
0.25	72.5	78.4	87	86.6
0.30	78	82.4	96	94.4
0.35	82.9	84.4	102	99
0.40	85	85.4	104	101

Tabular Data for Figure A-5

Appendix B

FLATHEAD COMPRESSION RATIO

ESTIMATING AND MEASURING FLATHEAD COMPRESSION RATIO

The compression ratio (CR) can be calculated as follows:

$$CR = \frac{CCFV + SV}{CCFV}$$

where,

CCFV = Total combustion chamber free volume at top dead center. See Table B-1 for values

SV = Swept Volume. Calculate using the formula below or see Table B-2 for values

$$= \frac{3.1416 \times Bore^2 \times Stroke}{4}$$

Features	Combustion Chamber Free Volume (cc) versus Head					
	8CM	**59A, 8BA, 8RT, EAC**	**EAB**	**Offy 425**	**Edelbrock & Offy 400**	**Offy 375**
Head combustion chamber	82	76	71	63	58	53
Valve c'bores in block with valves in place & piston deck clearance (0.02")	8	8	8	8	8	8
Piston dome (3/16" dome)	-13	-13	-13	-13	-13	-13
Head gasket (0.05" thick)	13	13	13	13	13	13
Relief (0.10" deep)	3	3	3	3	3	3
Typical Totals with above features	**93**	**87**	**82**	**74**	**69**	**64**

Table B-1: Combustion Chamber Free Volume (CCFV) in cubic centimeters for Selected Heads and Features. The values are within 5%

Bore (inches)	**Stroke** (inches)			
	3-3/4	4	4-1/8	4-1/4
3-1/16	453	483	498	513
3-3/16	490	523	539	556
3-5/16	530	565	582	600
3-3/8	550	586	605	623

Table B-2: Swept Volume (SV) in cubic centimeters Versus Bore and Stroke in inches

A sample calculation of compression ratio for an engine with a 3-5/16" bore and a 4" stroke, 8BA heads, pistons with a 3/16" dome, 0.020" piston deck clearance, 0.050" thick gasket and 0.100" deep relief is as follows using values from Tables B-1 and B-2:

$$CR = \frac{87 + 565}{87} = 7.5$$

If the engine has a 3-5/16" bore and a 4" stroke, 8BA heads, pistons with a 3/16" dome, 0.020" piston deck clearance, 0.050" thick gasket and no relief, we subtract out the volume of relief from the typical CCFV total as follows:

$$CR = \frac{(87 - 3) + 565}{87-3} = \frac{84 + 565}{84} = 7.7$$

We can see that "relieving" the engine reduces the compression ratio about 0.2 ratio points, i.e. from 7.7:1 to 7.5:1.

If in the above example, we substitute EAB heads for the 8BA variety, we get:

$$CR = \frac{(82 - 3) + 565}{82-3} = \frac{79 + 565}{79} = 8.2\,!$$

Another example for an engine with Offy 425 heads, a 3-5/16" bore and a 4" stroke, pistons with a 3/16" dome, 0.020" piston deck clearance, 0.050" thick gasket and 0.100" relief yields:

$$CR = \frac{74 + 565}{74} = 8.6$$

which is about maximum for this bore and stroke and still have reasonable flow characteristics.

You can estimate the effect on the free volume of combinations other than those in Table B-1. If you use flat top pistons, you don't subtract 13 cc for a piston dome. If you use a head gasket with a 0.030" compressed thickness, you add only 0.030/0.050 x 13 = 7.8 cc for gasket volume, and so on. You can also measure the volume in a completed engine, but this is messy and very difficult.

Appendix C

SUPERCHARGER KIT INSTALLATION INSTRUCTIONS

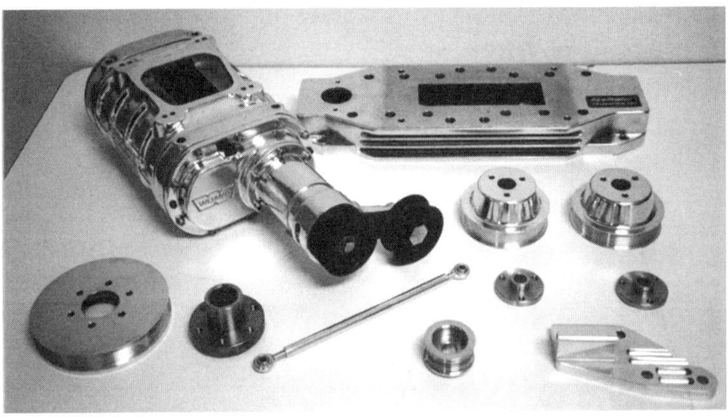

Figure C-1: Typical parts included in a Roadrunner Engineering Supercharger kit.

The following are typical installation Instructions for the Roadrunner Supercharger kit with a Weiand 142 for Ford Flatheads. Use instructions provided with each kit for actual assembly.

Preliminary Disassembly

1. Disconnect the battery. Drain the cooling system.

2. Remove existing intake manifold, carburetors, generator and fan, etc. Clean gasket surfaces. Clean existing threaded holes with proper tap.
3. Remove existing water pumps. Clean gasket surfaces. Clean threaded holes with tap.
4. Remove existing crankshaft hub/pulley assembly.

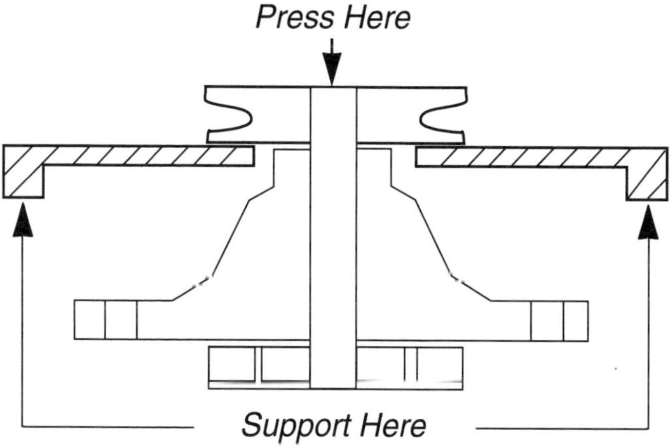

Figure C-2: Removing Stock Pulley From Water Pump

Water Pump Disassembly/Assembly for 1938-1948

NOTE: 1938-1948 water pumps are not recommended for serpentine belt drive unless they have been converted to a ball bearing system like the later pumps. As an alternate, the Roadrunner two belt "Nostalgia" supercharger kit for the early engines can be used with unmodified original pumps.

1. Press off existing pulleys being careful to apply pressing loads to pulley and shaft only. See Figure C-2.

2. Press on new pulley hubs to dimension shown in Figure C-3. Shims will be required between the hub and pump front face to maintain the proper impeller to rear pump surface clearance as shown in Figure C-3.

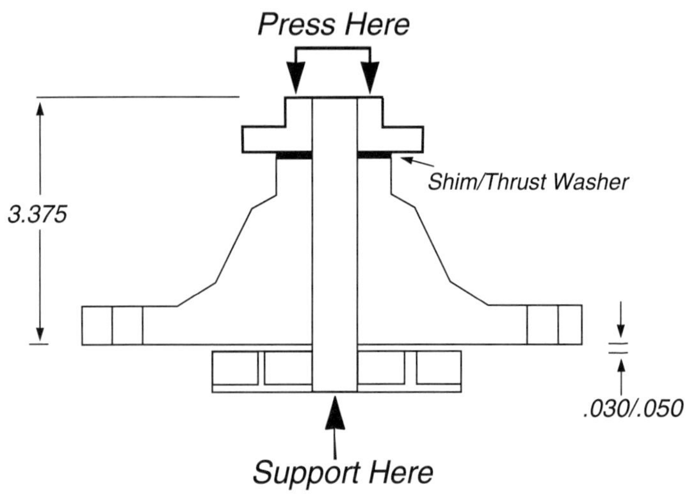

Figure C-3: 1938-1948 Water Pump Assembly with New Pulley Hub

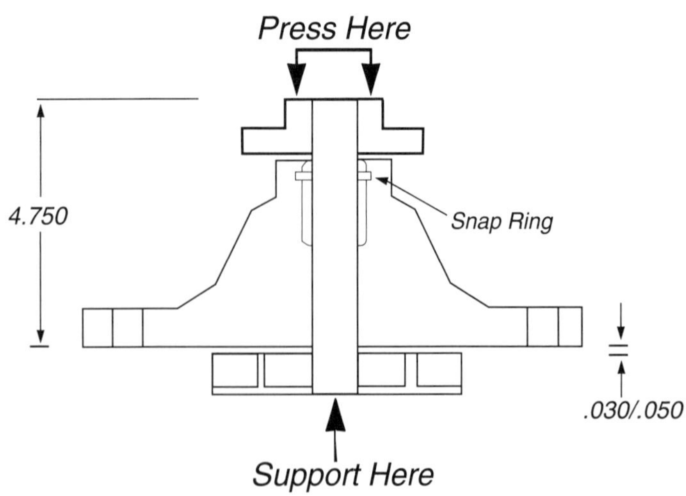

Figure C-4: 1949-1953 Water Pump Assembly

Water Pump Disassembly/Assembly for 1949-1953

1. Same as #1 above.
2. Press on new pulley hubs on both sides to dimension shown in Figure C-4. No shims required.

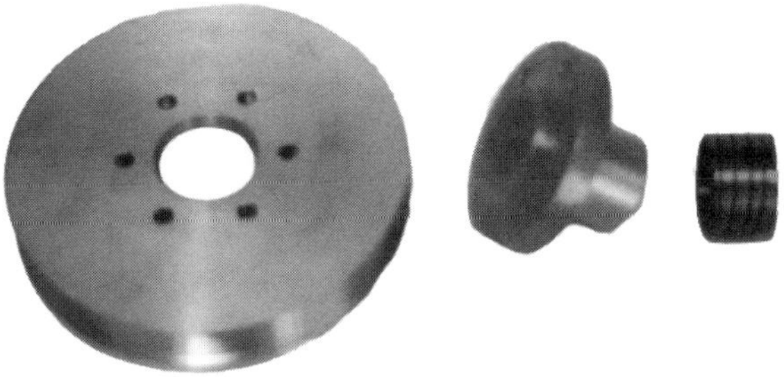

Figure C-5: Crankshaft Pulley, Hub and Spacer

Assembly

1. Install new crankshaft hub on engine. 1938-1948 engines will require spacer, Ford P/N 8BA-6362.
2. The crankshaft pulley, hub and spacer are shown in Figure C-5. This spacer will already be in place on newer engines. Torque 5/8"-18 retaining bolt to 135 ft-lbs.
3. Install water pumps using thread sealant (Permatex #14 or equivalent) on all 3/8"-16 retaining bolts and torque to 15-18 ft-lbs.
4. Install pulleys on water pump and crankshaft hubs. Torque 5/16"-24 retaining bolts to 12 ft-lbs. Use anti-seize lubricant (Permatex #133 or equivalent) on stainless fasteners where

used. Locate top dead center of #1 cylinder and add timing marks to crankshaft pulley before final installation.

5. Plug fuel pump push rod hole in engine. A 1/4 inch pipe plug is effective for this. The mechanical pump push rod will not be used. If this hole is left open, oil pressure will be lost.
6. *Optional. Drill and tap new intake manifold for boost/vacuum gauge line fitting. Installation of a boost gauge is highly recommended. Auto Meter makes a full line of these gauges. Install fitting.*
7. *Optional. Install blower manifold on engine. Temporarily check port alignment between engine and manifold with gasket in place. Match ports and gasket by grinding or cutting as necessary. This is an important step for high performance. It is particularly important that the block port does not significantly protrude into the blower manifold port. This condition is labeled as "bad" in Figure C-6. Severe turbulance is created in this case as air is pushed into the block causing high flow resistance.*

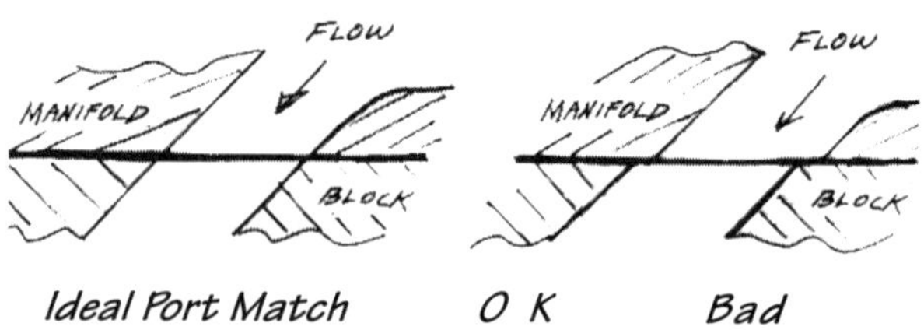

Figure C-6: Port Alignment Between Intake Manifold and Block

8. Install blower manifold and manifold gasket, then torque 3/8"-16 retaining bolts to 15-18 ft-lbs. 1949 and later engines will not have

as many mounting bolts as earlier engines (13 versus 17). Competition or high boost engines should consider installing the maximum number of bolts to enhance sealing. In this case, later blocks can be drilled and tapped appropriately.

9. Install blower assembly and gasket. Coat the gasket with talcum or baby powder to prevent sticking to manifold or supercharger. *Do not use any type of gasket sealant as this will void your warranty.* Make sure you have removed any tape residue with solvent after removing tape from bottom of the blower and take care not to let any foreign matter into the rotor housing. Torque the 7/16"-14 retaining bolts to 8-10 ft-lbs using a criss-cross pattern.

   While you are tightening the supercharger retaining bolts, turn the supercharger drive pulley by hand to make sure there is no bind. After tightening, remove the tape cover from the top of the supercharger and remove any residue.

10. Using a 0.010" feeler gauge, check the rotor to case clearance all along both of the upper edges of the case from front to back, inserting the gauge no more than 0.5" past the edges. If there is any bind against the feeler gauges, loosen the retaining bolts then retorque. Check lubricant levels in blower.

11. Obtain a *compact* alternator such as a Delco 121 series unit and shorten the mounting "foot" as required for proper pulley alignment when

installed on the bracket. The recommended alternator is designated by Lester (re-builder) number "7937." Alternators are available from Roadrunner in a variety of configurations and ready to bolt on.

12. Install alternator on the blower snout bracket. The alternator battery terminal (large one) goes to the battery (usually connected at the starter solenoid). For a 2-wire alternator such as the 7937 unit, the alternator "F" terminal is wired to the ignition switch "run" position.

13. Install tensioner rod to alternator and timing cover. 1948 and earlier engines will require removal of the vacuum brake on the distributor and plugging of the hole. The vacuum brake will not be used. See Chapter 5.

14. Install serpentine belt per Figure C-7. The required serpentine drive belt is best purchased and trial fitted locally because of all the combinations of pulleys and assemblies possible. These belts are commonly and inexpensively available at any auto supply in almost any size. The P/N's for these belts indicate the number of ribs, the rib design, and the length. As an example, if you have a street kit for the 1949-1953 flatheads with a 6" drive pulley and a 3.75" blower pulley, you may use a 6 rib, "K" series belt, 72.5 inches long. The AutoZone (Kelly Springfield) P/N is 725K6. A similar 71" long belt is P/N 710K6, and so on. The variable

position of the alternator bracket and the tensioner adjustment allows a belt to be fitted and tensioned easily.

Figure C-7: Serpentine Belt Configuration

15. Adjust belt tension. Proper belt tension can be obtained by feel and experience, however, a tension gauge is very helpful. A "Krikit II Pocket Gauge", Dayco P/N 93866, is handy and inexpensive. A belt tension of approximately 110-130 lbs. is a good place to start. Tension can be adjusted easily, usually by hand, using the tensioner rod. A jam nut is provided to lock the tensioner in place, but is usually not necessary. The tension is sufficient when the alternator pulley/fan can not be rotated easily by hand. Too little tension will allow the belt to slip under high boost. The symptoms are low boost and belt dust on the firewall. Too much tension will result in premature water pump bearing failure.

    A little experience will allow you to judge proper tension for average driving versus aggressive performance (high boost) driving. **Important – Recheck belt tension often.**

16. Install oil breather/oil filler assembly at rear of blower manifold. A stock unit may be used or an accessory item such as Roadrunner or Flathead Jack's P/N FJ-44 looks great. See Appendix G.

17. Install carburetor on blower using proper studs and nuts, and an appropriate gasket for the carburetor. A 600 cfm 4-barrel carburetor with vacuum operated secondaries will give good results in most applications. Make sure the carburetor throttle blades and linkage

operate smoothly without binding. Full throttle is essential for maximum performance. **A jammed throttle is extremely dangerous**.

A high capacity air cleaner is essential. The air cleaner top should be at least 1" above the carburetor choke tower.

18. See Chapter 5 and Appendix D for starting point on fuel systems and ignition systems as well as tuner's tips.
19. Connect carburetor linkage and fuel lines. Install electric fan. Refill cooling system and reconnect battery. Start engine, checking for leaks. Check and adjust idle and timing.
20. Go forth and kick those overheads around!

Appendix D

MOTORHEAD MART SPECIAL

The Motorhead Mart Special is a 1934 Tudor sedan which is used on both the street and strip. It is powered by a blown 276 cubic inch flathead Mercury. The 2660 pound car has run the quarter mile drags with best times of 100+ mph with 13.4 seconds elapsed times on gas, through the mufflers. The horsepower required to achieve these times was approximately 250 at Bakersfield and Phoenix.

Figure D-1: The Motorhead Mart Special at 1997 Goodguys "March Meet" in Bakersfield.

Under "standard" sea level conditions (60°F, 29.92 in. Hg), the current engine would make about 260 hp. The particulars for the car and engine are listed below.

Car

Body	1934 Ford Tudor 2660 lbs with the driver and fuel
Transmission	Ford C-4 with shift kit Stock converter, 1500 rpm stall
Rear End	Ford 9" with 4:11 gears Detroit Locker Pete & Jake's ladder bars
Tires	Mickey Thompson slicks (for racing) 28.5" diameter x 10" wide 7-10 lbs/in^2 pressure

Engine

Block	1949 Mercury with 0.125" over–sized bore No relief "Rolled" shortside (upper) radius Ported end exhaust outlets Offy main cap support, P/N 3408
Crankshaft	1949 Mercury 0.020/0.020 inches undersize Michigan 77 bearings
Rods	1949 Mercury Sanded, polished, & shot peened Michigan 77 bearings

Figure D-2: Motorhead Mart Special Motivation Compartment

Pistons	Arias forged aluminum, P/N 222379 Three ring Solid skirt Spiroloc pin retainers
Rings	Sealed Power, P/N 9760KX "Moly" upper ring, 1.5 mm
Cam	Isky 400 Jr., P/N 814000, 8BA Isky single valve springs, P/N 185-G No shims Johnson solid lifters
Valves	Manley "Pro Flo" stainless steel 1.6" diameter intake & exhaust Ford keepers

Heads	1949 Mercury "8CM" undercut in valve area Chamfered spark plug holes
Head Gasket	Fel Pro, P/N 1055/1056
Oil Pump	Mellings, M-19
Intake Manifold	Roadrunner Engineering, P/N 2001
Supercharger	Wieand 142 cubic inch or Camden 9" 3.75" pulley on blower Roadrunner drive system with 6.0" crankshaft pulley
Ignition	Mallory Dual Point distributor, P/N 2527501 Points gapped to 0.017" 4 degrees initial advance @ 750 rpm 20 degrees total advance @ 3000 rpm Mallory Voltmaster Coil Champion J6C (823) plugs with 0.035" gap
Carburetor	Holley 4150 carburetor - 600 cfm #70 primary jets #72 secondary jets #31 squirter Clear (plain) secondary spring 75 power valve
Fuel Pump	Holley "red" pump, P/N 12-801
Fuel Filter	Mallory Comp filter 140, P/N 3140

Appendix E

BLOWN FLATHEAD ALBUM

I heard a phrase once that goes like this - "If you don't know where you've been, you won't know where you're going." We all need a frame of reference to guide us and in this case, history can be fun!

Superchargers have been around for a long time. Mercedes Benz imported a supercharged car to the U.S. in 1924. Then there were the U.S. made supercharged Auburns, Cords, Dusenbergs, and Grahams. These supercharged cars were too expensive for most folks including most hot rodders. In Al Drake's book[1], he documented an interview with Vern Houle who claimed to be "the first guy in Los Angeles county with a blown flathead." Houle

McCulloch installed on 1940 Ford. The McCulloch Supercharger claimed a 38% horsepower increase for a cost of $124.50 including a dual exhaust system in 1939.

arrived in California in 1936 with a McCulloch supercharger on a 1935 Ford. The McCulloch was a centrifugal blower made in Milwaukee, WI, and 1936 was the first year of commercial manufacture.

For kits, besides the centrifugal McCulloch and the similar Frenzel built in Denver, there were sliding vane types such as the Arnolt, the Shurrock and the Judson, imported after WWII.

In the early 1950's, a Roots-type blower was imported from Italy under the Italemeccanica name, and a kit for the flathead Ford was available for about $500[2]. It worked well, but was expensive, about $5000 in current day dollars and not very reliable. It was later re-engineered in the U.S. and produced under the S.CO.T. name. These are good units and quite usable and collectable today.

The most common supercharger for competition flatheads was, and is, the GMC Roots-type blower produced for 2 cycle diesel engines. Of these, the 71 series GMC (3-71, 4-71, 6-71) are the most commonly used although the narrower, but less common 53 series is a better fit for a flathead. The GMC's are also used on street flatheads but are not a very good fit and there are no commercially available bolt-on kits for the GMC at this time.

Currently, the only complete blower kits being produced for the flathead are those produced by Roadrunner Engineering–my own company, and Motor City Flathead (see Appendix G). These kits use the popular B&M and Weiand Roots-type street blowers, both now produced by Holley.

1945: Dick Ford roadster

The Dick Ford roadster[1] utilized an English Arnolt vane-type blower producing 7.5 lbs/in^2 boost and was reputed to be one of the fastest streetrods of the time.

The earliest really successful supercharged flathead was Don Blair's "Goat," which used a Mercedes SSK blower on a 258 cubic inch Mercury, which turned 141 mph on the dry lakes.[3&4]

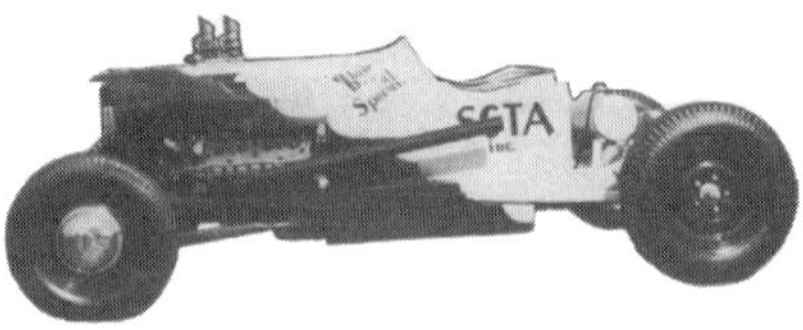

1946: Don Blair's "Goat"

1948: Barney Navarro's Hot Rod with Tom Beatty "support"

Barney Navarro[3] used a converted 3-71 GMC blower with 10 psi boost to produce 237 hp @5400 rpm from 239 cubic inches on alcohol. He was timed at 139 mph on the dry lakes in 1948. The engine was ported and relieved, had 8.8:1 aluminum heads, JE pistons, Winfield SU-1 cam, and four Stromberg 48 carburetors.

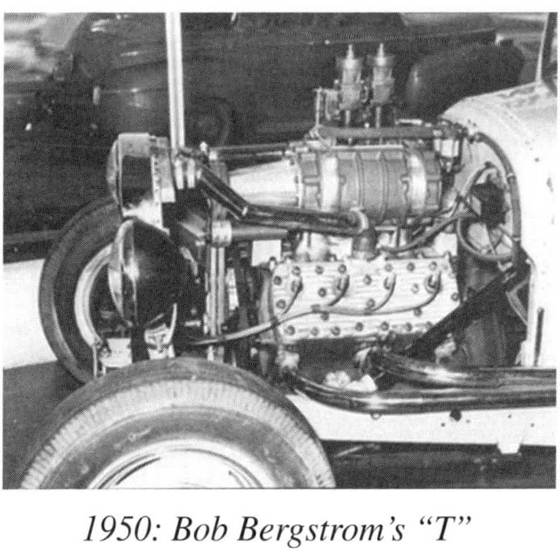

1950: Bob Bergstrom's "T"

Bob Bergstrom's [4] 1927 Model T roadster with a 3-71 GMC blower was one of the few early blown flathead street cars.

1951: Tom Cobb's Coupe/Roadster

Tom Cobb's coupe/roadster[5] is pictured with a GMC blower topped with an aircraft carburetor. Most of the time, he used 4 Stromberg carburetors. He ran 12-20 lbs/in^2 boost and took top "D" coupe (282-305 cubic inches with a blower) speed at 156 mph at the 1951 Bonneville nationals.

1952: Tom Beatty's Belly Tank

Tom Beatty's 3-71 GMC blown flathead[4&6] in a belly tank was the "most powerful flathead by far for many years" according to Mark Dees.

He turned 197 mph at Bonneville in 1952 and eventually went as fast as 199 mph at El Mirage and 211 mph at Bonneville in 1955. His 296 cubic inch engine was equipped with a Winfield SU-1A cam and Navarro heads, reportedly with 11.5:1 compression ratio (at least once!).

1955: Yates-Newmire Engine in Christman Dragster

Don Yate's and Bud Newmire's 3-71 GMC blown, 248 cubic inch, flathead[7&8] was billed as the "World's Most Powerful Flathead Ford V-8" in 1955. Using Hilborn injectors with 50% nitro, 17.5 lbs/in^2 boost (with 81% overdrive on blower) and 8:1 compression ratio, they got 435 hp at 5250 rpm! The cam was a Winfield SU-1A.

Yates and Newmire eventually reduced the nitro to 25% and the compression ratio to 6:1 for reliability and still got 425 hp. They turned 144 mph in 8.9 seconds in Christman's famous # 25 dragster at Santa Ana and this record stood for a very long time.

Yates-Newmire Engine

1957: Tony Nancy's "22 Jr"

This red "22 Jr" was Tony Nancy's last flathead powered competition car. Installed in a 1929 Ford, the destroked 256 cu. in. engine utilized a S.CO.T. blower, an Engle 300-B cam and running fuel, he turned the 1/4 mile at 128 mph in 10.47 seconds. "22 Jr" was featured in the December 1957 Hot Rod Magazine. The above picture is after its 1994 restoration.

1958: John Bradley with top mounted S.CO.T. supercharger

"Mr. Flathead," John Bradley[9], has run flathead dragsters for over forty years. Most have been unblown, but the pictures show two blown flathead dragsters. The GMC version turned 151 mph in 9.17 sec. The S.CO.T. turned 152 mph in 9.32 sec. Both were 296 cubic inch engines with Isky 404 cams.

Unblown, on 100% nitro, Bradley's best-to-date is 154 mph and 8.68 seconds elapsed time in the quarter mile at Sears Point in 1996.

1959: John Bradley's "Gene's Brake Shop Special" with front mounted GMC 4-71 blower

1960: Craig Breedlove's 1934 Coupe

None other than former jet car land speed record holder, Craig Breedlove, cut his teeth on a blown flathead powered 1934 Ford Coupe[10] in 1960. The 1948 Merc engine was bored 3/16" and destroked 1/8" for 258 cubic inches. Edelbrock 8:1 heads, a S.CO.T. blower with 4 Stromberg 48's, an Engle cam and JE pistons with Grant rings were used. The car turned 149 mph at the dry lakes resetting the then current record by 15 mph in it's class - B Coupe!

Craig Breedlove's Engine

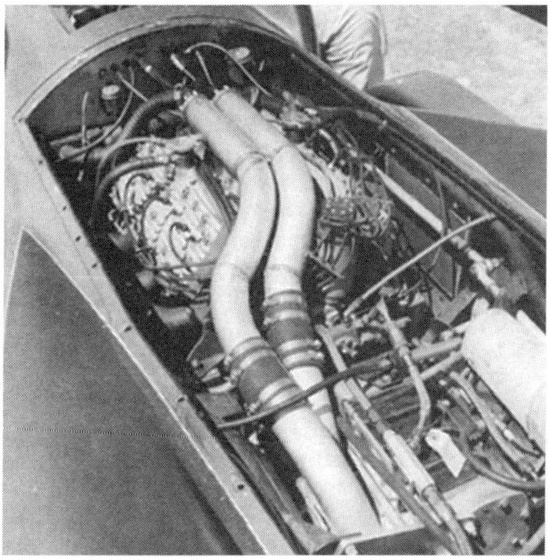

1971: Ed Harding's Streamliner

Ed Harding's "Redhead" streamliner[4] turned 229 mph at Bonneville in 1971 using a front driven 6-71 GMC blower on a 258 cubic inch flathead with a Winfield SU-1A cam and 1-5/8" valves.

Mark Dees[4] said that "if Tom Beatty's blown flathead wasn't the strongest side valve engine ever built for reasonably sustained running, Ed Harding's was."

1995: Stevens' and Houtz's Dragster

Jim Stevens' and Ed Houtz's 6-71 blown flathead dragster has been a strong competitor in recent years. Best to date is 158 mph and 8.83 seconds elapsed time in the quarter mile at Bakersfield in 1995.

1996: Costella-Murray Streamliner

Jack Costella's slippery streamliner, "Nebulous Theorem III" holds all current blown and unblown, gas and fuel, flathead top speed records at Bonneville.[11] The blown flathead versions run Don Murray's 3-5/16" x 4" (275 cu. in.) engine with Ross pistons, Isky 400 Jr. cam, 1-5/8 inch intake and exhaust valves, 8CM heads and a 0.100" block relief. The Turbo Power turbo-chargers put out 4 lb/in^2 boost drawing thru a 400 cfm Motorcraft 2-bbl carburetor. The car ran 235.795 mph (average) in the blown gas class and 235.647 in the blown fuel class (on gas). The unblown flathead version runs a 267 cu. in. engine owned by Jerry Helwig. Top speed with this engine was 220.109 (average) on fuel and 215.246 on gas.

Costella-Murray Engine

1998: Dave McCain's Dragster

Dave McCain's Dragster is consistently one of the strongest blown flathead dragsters. The car has turned a fastest time of 164 mph with an elapsed time of 8.18 seconds at Bakersfield, California in 1998 running 30% nitro. The engine has a 3-5/16 inch bore x 4-1/4 inch stroke (290 cubic inches), owner made heads, 1.72" intake valves, 1.6" exhaust valves, a Crane 450 cam with custom radiussed lifters and a 6-71 GMC blower (10% underdrive, 17psi boost). The transmission is a 4 speed Liberty and the car weighs in at 1515 lbs. without the driver.

1999: Rick Schnell's "Slider"

Rich Schnell's dragster is currently the quickest accelerating flathead known to the author. The 4-71 GMC blown flathead turned the quarter mile in 7.724 seconds at 173.73 mph in Wentzville, Missouri in 1999. The car weighs 1325 lbs. with driver. The engine runs with a 3.425" bore x 4.125" stroke (304 cu. in.), 1.75" intake valves, 1.5" exhaust valves, an Isky 433 cam and a 4-71 GMC blower with 10% overdrive. Heads are owner made. The transmission is a manually shifted Ford C-4. Estimated power output is approximately 600 hp on 40% nitro.

Footnotes and Photo Credits

[1] Drake, Albert. *Hot Rodder! From Lakes to Street.* Flatout Press, 1993.

[2] Francisco, Don. *Packing the Punch in Your Powerhouse.* Hot Rod, April 1952.

[3] Fisher, Bill. *California Bill's Hot Rod Manual.*Fred Fisher, 1949.

[4] *Technical History of the Racing Flathead, Part IV.* Early Ford V-8 Club, V-8 Times, Sept/Oct 1978.

[5] Fisher, Bill. *California Bill's Ford Speed Manual.* Floyd Clymer, 1952.

[6] Hot Rod 1954 Annual. No. 110. Trend Books, 1953.

[7] Navarro, Barney. *The Keeper of the Stable.* Rod & Custom Magazine. Quinn Publications, April 1955.

[8] Navarro, Barney. *Harnessing the Horses.* Rod & Custom Magazine. Quinn Publications, May 1995.

[9] John Bradley

[10] *Never Scrap a Coupe.* Hot Rod Magazine, Petersen Publishing, September 1960.

[11] Jack Costella

Front Cover: Corey Weintraub & Kathleen McCaughey

Back Cover: Bob Johnson Photography (805) 943-0855

NOTE:

Photos not labeled otherwise are by the author.

Appendix F

FLATHEAD READING

There have been many, many articles and books written on the flathead Ford engine. The observations and information in these writings are often different and even contradictory. This leaves the reader with the final responsibility for determining what is "right" for his/her application. The following list will make interesting reading for the flathead fan.

1. Bishop, Ron. *Rebuilding the Famous Ford Flathead*. Lab Books, Inc., 1981.
2. Ceridono, Ron, editor. *The Complete Ford Flathead V8 Engine Manual*. Tex Smith Publishing and CarTech, Inc., 1995.
3. Davidson, Mike. *Flathead Fever*, Second Edition. Graffiti Press, 1996.
4. Dees, Mark. *A Technical History of the Racing Flathead*. Volume 15, Numbers 1, 2, 4, and 5. Early Ford V-8 Club of America "V-8 Times," 1978. Partially reprinted from Rod & Custom Magazine, Volume 20, 1973.

5. Fisher, "California Bill." *Ford Speed Manual.* Originally published by Floyd Clymer, 1949 and 1952. Reprinted by Fisher Books, 1995.
6. Ganahl, Pat, *A Do-It-Yourself Guide to Street Supercharging,* S-A Design Books, 1984.
7. Ganahl, Pat, *Flatheads Forever.* Volume 23, Numbers 2, 3 and 4. Rod & Custom Magazine, 1989.
8. Hagens, Jerry. *Fullrace Flathead V-8.* Jerry Hagens, 1981.
9. Holleran, Ron. *Nostalgia - Rebuilding and Modifying the Flathead Ford V-8.* Ron Holleran, 1992.
10. Huntington, Roger. *How to Hop Up Ford & Mercury V-8 Engines.* Originally published by Clymer Books, 1951. Reprinted by Fisher Books, 1997.
11. Huntington, Roger. *Souping the Stock Engine.* Originally published by Floyd Clymer, 1950. Reprinted by Fisher Books, 1997.
12. Oddo, Frank. *A Flathead for the Nineties.* Volume 23, Numbers 2 through 12. Street Rodder Magazine, 1994.
13. Oddo, Frank, *Ford Flathead V-8 Builders Handbook 1932-1953.* Fisher Books, 1997.
14. Pagé, Victor. *Early Ford V-8 Ford Service Manual.* Clymer Publications, various editions through 1979.
15. Sinclair, Bill. *Thunder Road Flathead.* Thunder Road Motor Works, Inc., 1991.

16. *How to Build Hot Rods.* Fawcett Book, Number 156. Fawcett Publications, 1952.
17. Motor's Auto Repair Manual. Motor Magazine, various editions through 1962.
18. Ford Motor Co. *Ford and Mercury Shop Manual (Cars & Trucks)*, 1939-1948. Canadian factory manual. Reprinted by Crank-En Hope Publications.
19. Ford Motor Co. *Ford Passenger Car Shop Manual*, 1949-1951. Factory Manual. Reprinted by Various Publishers.
20. Ford Motor Co. *Ford Passenger Car Shop Manual,* 1952 with Supplements for 1953 and 1954. Factory manual reprinted by Lincoln Publishing Co., 1997.
21. Ricardo, Harry R. *The High Speed Internal Combustion Engine*, Blackie and Son Limited, London, 1953.

NOTE

Many of the above titles and others, including this book, are available from your local bookstore or:

Classic Motorbooks

P. O. Box 1

Osceola, WI 54020-0001

Phone: ***(800) 826-6600***

website: www.motorbooks.com

Appendix G

FLATHEAD BUILDER'S TIPS and SUPPLIERS

RUST REMOVAL

Acid cleaning is recommended for heavy rust removal from steel or iron parts. Because of the hazards involved, this should be done professionally where facilities are available. If such facilities are not readily available, acid cleaning can be effectively accomplished with a solution of one part muriatic acid to three parts water. This acid is commonly and inexpensively available at home and swimming pool supply stores. A caustic cleaning is necessary to remove grease and oil prior to acid cleaning.

Cleaning of flathead cooling passages can be done without immersion by installing iron heads and filling an upright (bell housing surface down) block assembly with the acid solution. Rust removal may take hours or days depending on the amount of scale present. You can practice on small rusty parts to get a feel for the process. Be sure and follow all appropriate safety procedures. A *Material Safety Data Sheet* will be available from the acid supplier. The acid cleaning should be followed by thorough rinsing and another caustic dip to neutralize the acid.

THREADED JOINTS

All threads should be tapped or chased and cleaned prior to assembly. A calibrated torque wrench should be used whenever possible for tightening threaded joints and is mandatory for head, rod cap and main cap installation. Where sealing and galling are not concerns, the joints should be lubricated with engine oil prior to assembly for the most consistent torque readings. The tightening should be done in two or more stages and in a criss-cross pattern or other pattern (e.g. head bolts/nuts) specified by the factory or equipment supplier.

An anti-seize compound of some type is recommended for stainless steel, chrome plated, or aluminum threads to prevent galling. Also, care should be taken not to over tighten these threads since most are equivalent to grade 2 or less. Permatex thread sealant, P/N 14A, provides anti-seize, thread sealing and some locking capability and is an excellent choice for head bolts or studs. Medium strength Loctite (blue) thread locking compound provides some anti-seize protection and sealing qualitities while providing adequate thread locking where lock washers are not used.

SUPPPLIERS

There are many suppliers of flathead services and parts, both stock and high performance. The following list is just a sampling to get you started.

Car Products *(Parts)*
P. O. Box 96
Paragonah, UT 84760
Phone: (435) 477-8213

Dennis Carpenter Reproductions *(Parts, **new** and rebuilt engines)*
P. O. Box 26398
Charlotte, NC 28221-6398
Phone: (800) 476-9653

Flathead Jack *(Parts)*
1516 Third Ave
Walnut Creek, CA 94596
Phone: (888) 993-2233
Fax: (510) 932-1851

Hot Rod & Custom Supply *(Parts & Service)*
1304 S.E. 10th Street
Cape Coral, FL 33990
Phone: (941) 574-7744

Mike's Sidevalve Supply *(Parts)*
51 First Avenue
St. Peters, Adelaide 5069
Australia
Fax: 61 8 8362 1515

Motor City Flathead *(Parts & Service)*
14245 Stowell Road
Dundee, MI 48131
Phone: (734) 529-3363

Red's Headers *(Parts & Service)*
22950 Bednar Lane
Fort Bragg, NC 95437
Phone: (707) 964-7733

Roadrunner Engineering *(Flathead Engine Analysis and Supercharger Kits)*
P.O. Box 53296
Albuquerque, NM 87153-3296
Phone/Fax: (505) 268-6768

Patrick's *(Parts)*
P. O. Box 10648
Casa Grande, AZ 85230
Phone: (520) 836-1117

Speedway Motors *(Parts)*
P. O. Box 81906
Lincoln, NE 68501
Phone: (402) 323-3200
Fax: (800) 736-3733

Vintage Ford Parts *(Parts)*
11318 Beach Blvd.
Stanton, CA 90680
Phone: (714) 894-5464
Fax: (714) 894-4563